Published by: AoPS Incorporated
                     15330 Avenue of Science
                     San Diego, CA 92128
                     info@BeastAcademy.com

ISBN: 978-1-934124-65-9

Beast Academy is a registered trademark of AoPS Incorporated.

Written by Jason Batterson, Shannon Rogers, and Kyle Guillet
Book Design by Lisa T. Phan
Illustrations by Erich Owen
Grayscales by Greta Selman

Visit the Beast Academy website at BeastAcademy.com.
Visit the Art of Problem Solving website at artofproblemsolving.com.
Printed in the United States of America.
2022 Printing.

Contents:

This is Practice Book 5C in a four-book series.

5A
• 3D Solids
• Integers
• Expressions & Equations

5B
• Statistics
• Factors & Multiples
• Fractions

5C
• Sequences
• Ratios & Rates
• Decimals

5D
• Percents
• Square Roots
• Exponents

For more resources and information, visit BeastAcademy.com.

This is Beast Academy Practice Book 5C.

Each chapter of this Practice book corresponds to a chapter from Beast Academy Guide 5C.

MATH PRACTICE 5C

MATH GUIDE 5C

The first page of each chapter includes a recommended sequence for the Guide and Practice books.

You may also read the entire chapter in the Guide before beginning the Practice chapter.

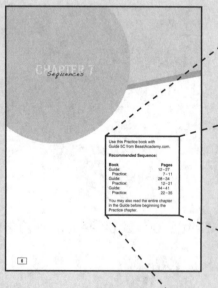

Use this Practice book with Guide 5C from BeastAcademy.com.

Recommended Sequence:

Book	Pages
Guide:	12 – 27
Practice:	7 – 11
Guide:	28 – 34
Practice:	12 – 21
Guide:	34 – 41
Practice:	22 – 35

You may also read the entire chapter in the Guide before beginning the Practice chapter.

Some problems in this book are very challenging. These problems are marked with a ★. The hardest problems have two stars!

Every problem marked with a ★ has a *hint!*

Hints for the starred problems begin on page 108.

Other problems are marked with a ✎. For these problems, you should write an explanation for your answer.

54. ★

55. ✎

42 Guide Pages: 39-43

Some pages direct you to related pages from the Guide.

None of the problems in this book require the use of a calculator.

Solutions are in the back, starting on page 112.

A complete explanation is given for every problem!

CHAPTER 7
Sequences

Use this Practice book with
Guide 5C from BeastAcademy.com.

Recommended Sequence:

Book	Pages:
Guide:	12-27
Practice:	7-11
Guide:	28-34
Practice:	12-21
Guide:	35-41
Practice:	22-35

You may also read the entire chapter
in the Guide before beginning the
Practice chapter.

Math beasts are always looking for patterns!

PRACTICE | Circle the figure that best completes each pattern below.

1.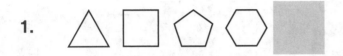

2.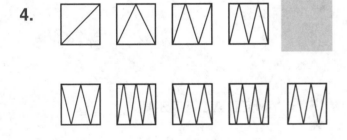

3.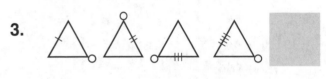

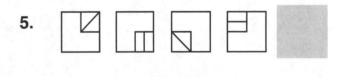

4.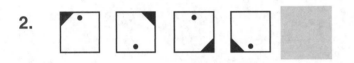

5.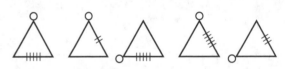

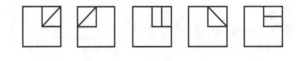

6.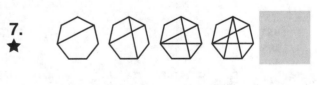

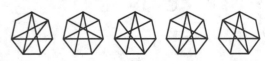

7. ★

What Comes Next?

Some of these are very difficult.

If you don't get them, you can always move ahead and come back later!

PRACTICE | Write the best choice for the next character in each pattern below.

8. A c E g I k _____

9. S M T W T F _____

10. A E I M Q U Y C G _____

11. Z ⅄ X Ϻ V ∩ _____

12. O T T F F S S E _____

13. A B D G K P _____

14. ★ M V E M J S U _____

15. ★ Y M E D A C A T S A E _____

16. ★★ A B E C F D H G I J K _____

17. ★★ B C D E G P T _____

A **sequence** is a list of terms, usually numbers, that often follow a pattern.

EXAMPLE | Find the next term in the sequence below.

$$3, 6, 12, 24, 48, \underline{\quad}$$

In this sequence, each term is double the value of the term before it. So, the next term in the sequence is $48 \cdot 2 = \mathbf{96}$.

$$3, 6, 12, 24, 48, \underline{\ 96\ }$$

PRACTICE | Write the next three terms that best fit each sequence below.

18. 1, 2, 4, 7, 11, 16, _____, _____, _____

19. 46, 40, 32, 22, 10, _____, _____, _____

20. 640, -320, 160, -80, 40, _____, _____, _____

21. 10, 20, 14, 28, 22, 44, 38, _____, _____, _____

22. 1, 2, 6, 24, 120, _____, _____, _____

23. ★ $\frac{1}{2}$, 2, $\frac{9}{2}$, 8, $\frac{25}{2}$, 18, _____, _____, _____

EXAMPLE | If the pattern below continues, what will be the 50th term in the sequence?

$$4,\ 8,\ 12,\ 16,\ 20,\ ...$$

To find the 50th term, we look for a relationship between each term and its position in the sequence.

Position: 1st 2nd 3rd 4th 5th
Term: 4, 8, 12, 16, 20

The 1st term in the sequence is $1 \cdot 4 = 4$.
The 2nd term in the sequence is $2 \cdot 4 = 8$.
The 3rd term in the sequence is $3 \cdot 4 = 12$, and so on.

Each term is 4 times its position number.
So, the 50th term in the sequence is $50 \cdot 4 = \mathbf{200}$.

Position: 1st 2nd 3rd 4th 5th ... 50th
Term: 4, 8, 12, 16, 20, ..., 200 $\big){\times}4$

> The three dots (...) at the end of a sequence tell us that it continues forever.
> Sequences that continue forever are called *infinite sequences*.

PRACTICE | In each sequence below, continue the pattern to find the missing terms listed.

24. 19, 20, 21, 22, 23, ...

24.   15th term: _____

   50th term: _____

25. 1, 8, 27, 64, 125, ...

25.   9th term: _____

   20th term: _____

26. 3, -6, 9, -12, 15, ...

26.   25th term: _____

   100th term: _____

27. 0, 7, 0, 14, 0, 21, ...

27.   99th term: _____

   40th term: _____

28. ★ 1, 11, 111, 11, 1, 11, 111, 11, 1, ...

28.   80th term: _____

   51st term: _____

EXAMPLE | Continuing the pattern below, find the 10th and 25th terms in the sequence. Then, write an expression for the nth term of the sequence.

$$\frac{1}{3}, \ \frac{1}{4}, \ \frac{1}{5}, \ \frac{1}{6}, \ \frac{1}{7}, \ \cdots$$

Each term in the sequence is a unit fraction with a denominator that is 2 more than the term's position number.

So, the 10th term is $\frac{1}{10+2} = \frac{1}{12}$, the 25th term is $\frac{1}{25+2} = \frac{1}{27}$, and the nth term is $\frac{1}{n+2}$.

Position:	1st	2nd	3rd	4th	5th	$\cdots$	10th	$\cdots$	25th	$\cdots$	n^{th}
Term:	$\frac{1}{3},$	$\frac{1}{4},$	$\frac{1}{5},$	$\frac{1}{6},$	$\frac{1}{7},$	$\cdots$	$\frac{1}{12},$	$\cdots$	$\frac{1}{27},$	$\cdots$	$\frac{1}{n+2}$

PRACTICE | For each sequence below, fill in the terms in the positions listed beneath the blanks, and write an expression for the nth term of the sequence.

29. 2, 4, 6, 8, 10, ..., _____ , ..., _____ , ..., _____
 25th 50th n^{th}

30. 1, 3, 5, 7, 9, ..., _____ , ..., _____ , ..., _____
 30th 75th n^{th}

31. 2, 4, 8, 16, 32, _____ , ..., _____ , ..., _____
 6th 10th n^{th}

32. 1, 4, 9, 16, 25, ..., _____ , ..., _____ , ..., _____
 10th 25th n^{th}

33. ★ -1, 1, -1, 1, -1, 1, ..., _____ , ..., _____ , ..., _____
 15th 40th n^{th}

34. ★ 2, 1, $\frac{2}{3}$, $\frac{1}{2}$, $\frac{2}{5}$, $\frac{1}{3}$, $\frac{2}{7}$, $\frac{1}{4}$, ..., _____ , ..., _____ , ..., _____
 20th 35th n^{th}

In an **arithmetic sequence**, the same amount is always added to get from one term to the next.

The amount that is added to get to each next term is called the **common difference**.

EXAMPLE | Fill in the blanks to complete the arithmetic sequence below.

___, 19, ___, ___, ___, 67, ___

Arithmetic sequences are really just skip-counting patterns!

We begin by finding the common difference. To get from 19 to 67 in this sequence, we add the common difference 4 times.

___, 19, ___, ___, ___, 67, ___

Adding the common difference 4 times adds a total of 67 − 19 = 48. So, the common difference is 48 ÷ 4 = 12. We use this to find the missing terms, as shown.

$$\underset{7}{}, 19, \underset{31}{}, \underset{43}{}, \underset{55}{}, 67, \underset{79}{}$$

+12 +12 +12 +12 +12 +12

PRACTICE | Find the common difference for each arithmetic sequence below.

35. 7, 16, 25, 34, 43, ...

36. -33, -25, -17, -9, -1, ...

37. 29, 26, 23, 20, 17, ...

38. __, 21, __, 35, __, ...

39. 74, __, __, 41, __, ...

40. 30, __, __, __, __, $32\frac{1}{2}$, ...

35. _____

36. _____

37. _____

38. _____

39. _____

40. _____

PRACTICE | Fill in the blanks to complete each arithmetic sequence below.

41. 19, _____, _____, 64, _____, _____, 109

42. 98, _____, 112, _____, _____, _____, 140

43. _____, _____, _____, _____, 32, $30\frac{1}{2}$, _____

44. 10, _____, _____, _____, _____, 22, _____

PRACTICE | Answer each question below.

45. What is the common difference of an arithmetic sequence whose first term is 25 and whose tenth term is 115?

45. _____

46. What is the common difference of an arithmetic sequence whose 23rd term is $\frac{1}{3}$ and whose 25th term is $\frac{1}{2}$?

46. _____

47. An arithmetic sequence has 10th term 4 and 30th term 68. What is the 20th term of the sequence?

47. _____

48. ★ An arithmetic sequence has five terms. The first term is 40, and the sum of all five terms is 80. What is the common difference?

48. _____

EXAMPLE | What is the 50th term of the arithmetic sequence below?

$$2, \; 5, \; 8, \; 11, \; 14, \; 17, \; ...$$

The first term of the arithmetic sequence is 2 and the common difference is 3.

$$\begin{array}{ccccc} +3 & +3 & +3 & +3 & +3 \end{array}$$
$$2, \; 5, \; 8, \; 11, \; 14, \; 17, \; ...$$

To get to the 2nd term, we add 1 three to 2.
To get to the 3rd term, we add 2 threes to 2.
To get to the 4th term, we add 3 threes to 2.

To get to the 50th term, we add 49 threes to 2.
So, the 50th term is $2+49(3) = 2+147 =$ **149**.

PRACTICE | Find the value of the missing term listed for each arithmetic sequence below.

49. 15, 19, 23, 27, 31, ..., $\underline{\qquad}_{10^{th}}$

50. -11, -6, -1, 4, 9, ..., $\underline{\qquad}_{40^{th}}$

51. 5, -2, -9, -16, -23, ..., $\underline{\qquad}_{15^{th}}$

52. -29, -19, -9, 1, 11, ..., $\underline{\qquad}_{100^{th}}$

PRACTICE | Answer each question below.

53. What is the 13th term of an arithmetic sequence whose first term is 9 and whose common difference is 8?

53. _____

54. What is the first term of an arithmetic sequence whose 100th term is 40 and whose common difference is $\frac{1}{3}$?

54. _____

55. The 12th and 15th terms of an arithmetic sequence are 85 and 106. What is the first term of the sequence?

55. _____

EXAMPLE | Write an expression for the nth term of the arithmetic sequence below.

$$-1, \ 4, \ 9, \ 14, \ 19, \ 24, \ ...$$

The first term of the sequence is -1 and the common difference is 5.

$$-1, \ 4, \ 9, \ 14, \ 19, \ 24, \ ...$$

To get the 2nd term, we add 1 five to -1.
To get the 3rd term, we add 2 fives to -1.
To get the 4th term, we add 3 fives to -1.

To get the nth term, we add $(n-1)$ fives to -1.

So, the nth term is $-1+(n-1)5$. Distributing the 5 and simplifying gives

$$-1+(n-1)5 = -1+5n-5$$
$$= 5n-6.$$

> You can check your answer by plugging in values for n!
>
> For example, we know the 3rd term is 9. So, we can plug in $n=3$ to get $5(3)-6=9.$ ✓

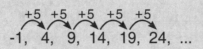

PRACTICE | Write a simplified expression for the nth term of each arithmetic sequence below.

56. 18, 24, 30, 36, ..., _____
n^{th}

57. 4, 19, 34, 49, ..., _____
n^{th}

58. -13, -5, 3, ..., _____
n^{th}

59. $\frac{9}{4}, \frac{5}{2}, \frac{11}{4}, ...,$ _____
n^{th}

PRACTICE | Answer each question below. Simplify all expressions.

60. An arithmetic sequence has first term a and common difference 3. Write an expression for the 20th term of the sequence.

60. _____

61. An arithmetic sequence has first term 6 and common difference d. Write an expression for the 101st term of the sequence.

61. _____

62. ★ In an arithmetic sequence, the 1st term is 20, the 2nd term is 32, and the kth term is 500. What is k?

62. $k =$ _____

In a **Sequence Path** puzzle, the goal is to draw one or more paths that connect at least three numbers in an order that forms an arithmetic sequence.

Every number must be part of a path, and no two paths can cross the same square.

16			19
	11	15	
		21	
23			

EXAMPLE | Solve the Sequence Path puzzle to the right.

Writing the numbers in the grid in order from least to greatest, we have 11, 15, 16, 19, 21, 23.

These six numbers do not form a single arithmetic sequence. Each arithmetic sequence has 3 or more terms, so we must have two 3-term arithmetic sequences.

Since 11 is the smallest number in the grid, it must be one end of a path. There are only two arithmetic sequences we can make that begin with 11 and include 3 terms:

11, 15, 16, 19, 21, 23

11, 15, 16, 19, 21, 23

Only the sequence 11-16-21 leaves numbers that also form an arithmetic sequence: 15-19-23. So, we connect the numbers in these sequences in order, as shown below.

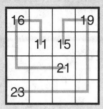

PRACTICE | Solve each Sequence Path puzzle below.

63.

	33	19	
		26	
	40	12	

64.

21	15		9
12			
		3	7
17			

65.

	6		3	10
		4	12	
			21	
			8	

66.

12	34		
		30	18
26	24		

67.

	61	57	
	67	66	
56			62

68.

22			26
38			30
32			34
24			40

PRACTICE | Solve each Sequence Path puzzle below.

69.

	34	43	
27	72		
		53	59

70.

74	83		80	86
		89		
77				92

71.

22				46
	34	52		
	40	58		
10				28

72.

	53			
			14	
		92		66
40	79			27

73. ★

	17	25	4	9	
15					
			12		
		13			
				5	20

74. ★ ★

33				51	21
	75	9			
			63		
		57			
			39	69	
81	27				45

75. ★

49	50				46	47
			45			
			43			
		44				
		60		42	48	
					40	41

76. ★ ★

		45				
						9
	57	53				
			5		61	
	49		13			
				41	1	
	17					
				65		

PRACTICE | Answer each question below.

77. Fill in the missing entries for each of the arithmetic sequences below.

a. 5, 35, _____, _____, _____, _____, _____, ...

b. 5, _____, 35, _____, _____, _____, _____, ...

c. 5, _____, _____, 35, _____, _____, _____, ...

d. 5, _____, _____, _____, _____, 35, _____, ...

78. Fill in the missing entries for each of the arithmetic sequences below.

a. _____, _____, _____, 27, _____, 39, _____, ...

b. _____, _____, 27, _____, _____, 39, _____, ...

c. _____, 27, _____, _____, _____, 39, _____, ...

d. 27, _____, _____, _____, _____, _____, 39, ...

79. How many common differences are possible for an arithmetic sequence of integers that starts with 9 and includes 33?

79. _____

80. How many common differences are possible for an arithmetic sequence of integers that starts with 15 and includes 46?

80. _____

81. ★ An arithmetic sequence of integers starts with 13, ends with 37, and includes exactly one other prime. What is the other prime in the sequence?

81. _____

PRACTICE | Answer each question below.

82. What is the greatest possible common difference of an arithmetic sequence of integers that includes the terms 27, 34, and 49?

82. _____

83. What is the largest term that can appear in an arithmetic sequence of 10 integers that includes 20, 32, and 50?

83. _____

84. ★ What is the smallest positive integer that could appear in an arithmetic sequence of non-consecutive integers that includes 23 and 58?

84. _____

85. ★ A 7-term arithmetic sequence of positive integers includes the terms 33, 53, and 73. Circle each number below that could also be a term in the sequence.

3   13   28   38   63   153

86. ★ What is the common difference of an eight-term arithmetic sequence of integers with three numbers in the 30's, two numbers in the 40's, and three numbers in the 50's?

86. _____

87. ★ Fill each blank below with a digit to create an arithmetic sequence of 2-digit integers.

__4, 3__, __4, __9, 8__, ___

In a **Cross-Sequence** puzzle, the goal is to fill every empty square with a positive integer so that the numbers in each row and each column can be arranged in some order to form an arithmetic sequence.

EXAMPLE | Fill the empty squares to complete the Cross-Sequence puzzle to the right.

	6	22
	13	16
18		

The center column contains a 6 and a 13. To create an arithmetic sequence that includes 6 and 13, we have ($\boxed{-1}$, 6, 13) or (6, $\boxed{9.5}$, 13) or (6, 13, $\boxed{20}$). Since each square must contain a positive integer, the bottom-center square must contain **20**. We only consider entries that are positive integers for the remaining squares.

	6	22
	13	16
18	**20**	

The bottom row is ($\boxed{16}$, 18, 20) or (18, $\boxed{19}$, 20) or (18, 20, $\boxed{22}$). Only placing **19** in the bottom-right square also creates an arithmetic sequence in the right column: 16, 19, 22.

	6	22
	13	16
18	**20**	**19**

Next, the middle row is ($\boxed{10}$, 13, 16) or (13, 16, $\boxed{19}$). If we fill the middle-left square with a 19, then the top-left square must be filled with a 17 or a 20 to create an arithmetic sequence in the left column. Neither of these creates an arithmetic sequence in the top row. So, the middle-left square is **10**.

17 or 20 ↓

	6	22
19	13	16
18	**20**	**19**

	6	22
10	13	16
18	**20**	**19**

Finally, only a **14** in the top-left square creates an arithmetic sequence in both the left column and the top row.

14	6	22
10	13	16
18	**20**	**19**

PRACTICE | Fill the empty squares to complete each Cross-Sequence puzzle below.

88.

1	4	
5		8
	6	9

89.

2		5
7	9	
		8

90.

	25	
3		16
1		14

91.

	60	
1	32	
99		77

92.

17		22
	36	
34		23

93.

6		15
11	33	
		8

Some of these puzzles are hard. If you get stumped by one, it's okay to come back to it later!

You do not need to create arithmetic sequences in rows or columns that are interrupted by black squares.

PRACTICE | Fill the empty squares to complete each Cross-Sequence puzzle below.

94.

	15	
	2	
5		
79		20

95.

7		14
22		
		18
	84	

96.

	19	
55		11
	5	
35		9

97.

		72	34
8			32
1		91	

98.

		46	
9			45
19			
		7	23

99.

	4	5	
			13
30			
	57		18

100.

	2		
	17		
1		4	13
		15	
	22		

101. ★

1	16			6
		17		45
22			13	
15				
	30	33		

102. ★★

	43		17	
22				26
35				28
	16		11	

103. ★

	38	8	44	14		
29						90
		33	43	53	63	

104. ★

	10		65		54	
28		75		77		3
	22				87	

The number of dots in each figure below represents a **triangular number**. The first four triangular numbers are 1, 3, 6, and 10.

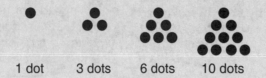

1 dot   3 dots   6 dots   10 dots

EXAMPLE | What is the 50th triangular number?

The 50th triangular number is given by the number of dots in the triangular pattern with 50 rows.

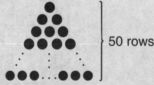

50 rows

This is equal to the sum $1+2+3+\cdots+48+49+50$.

To compute this sum, we add two copies of it, with one copy written in reverse.

$$
\begin{array}{r}
1 + 2 + 3 + \cdots + 48 + 49 + 50 \\
50 + 49 + 48 + \cdots + 3 + 2 + 1 \\
\hline
51 + 51 + 51 + \cdots + 51 + 51 + 51
\end{array}
$$

This gives 50 pairs of numbers that each sum to 51. So, the sum of all 50 pairs is $50 \cdot 51 = 2{,}550$. However, this is the total of **two** copies of the sum we wish to compute. So, the value of **one** copy is $2{,}550 \div 2 = 1{,}275$.

So, the 50th triangular number is **1,275**.

PRACTICE | Answer each question below.

105. What is the 40th triangular number?

105. _____

106. The nth triangular number is given by $1+2+3+\cdots+(n-2)+(n-1)+n$. Use the strategy shown in the example to write a simpler expression for the nth triangular number.

106. _____

107. Compute $2+4+6+8+10+12+14+16+18+20+22+24$.

107. _____

PRACTICE | Compute each sum below.

108. $88+80+72+64+56+48+40+32+24+16+8$

108. _____

109. $6+12+18+\cdots+588+594+600$

109. _____

110. $\frac{4}{7}+1\frac{1}{7}+1\frac{5}{7}+2\frac{2}{7}+\cdots+11\frac{3}{7}$

110. _____

111. $51+52+53+\cdots+98+99+100$

111. _____

PRACTICE | Answer each question below.

112. Ariana writes a sequence whose n^{th} term is found by subtracting the nth triangular number from the nth perfect square (where 1 is the first perfect square). What is the 25th term in Ariana's sequence?

112. _____

113. ★ Kropple subtracts the 997th triangular number from the 1,002nd triangular number. What is Kropple's result?

113. _____

114. ★ The kth triangular number is 465. What is k?

114. $k=$ _____

An **arithmetic series** is a sum of terms that form an arithmetic sequence. For example, 3+6+9+12 is an arithmetic series with a sum of 30. The sums that give triangular numbers $(1+2+3+\cdots+n)$ are also examples of arithmetic series.

EXAMPLE | Compute the sum of the following arithmetic series.

$$50+52+54+56+58+60+62+64+66+68$$

We add two copies of this series, with one copy written in reverse.

$$
\begin{array}{c}
50 + 52 + 54 + 56 + 58 + 60 + 62 + 64 + 66 + 68 \\
68 + 66 + 64 + 62 + 60 + 58 + 56 + 54 + 52 + 50 \\
\hline
118+118+118+118+118+118+118+118+118+118
\end{array}
$$

This gives 10 pairs of numbers that each sum to 118. The sum of all 10 pairs is $10 \cdot 118 = 1{,}180$. However, this is the sum of **two** copies of the series we wish to compute. So, **one** copy equals $1{,}180 \div 2 = \textbf{590}$.

— *or* —

To compute the sum, we find the average of the terms and multiply by the number of terms.

In any arithmetic sequence, the average of the terms is equal to the median. For example, the terms in this series balance around the median of 59.

50, 52, 54, 56, 58, 60, 62, 64, 66, 68

For every term that is greater than the median, there is a term that is less than the median by the same amount. So, the average of the terms is also 59.

The sum of 10 terms with an average of 59 is $10 \cdot 59 = \textbf{590}$.

PRACTICE | Compute the sum of each arithmetic series below.

115. $4+5+6+7+8+9+10+11+12+13+14+15+16$

115. _____

116. $20+30+40+50+60+70+80+90$

116. _____

117. $10+21+32+43+54+65+76+87+98$

117. _____

118. $(-6)+(-5)+(-4)+(-3)+(-2)+(-1)+0+1+2+3+4+5+6+7+8$

118. _____

119. $11+17+23+29+35+41+47+53+59+65$

119. _____

Review counting the number of terms in a list in the Counting chapter of Beast Academy 4B.

PRACTICE | Answer each question below about the following sequence.

9, 13, 17, 21, 25, ..., 45, 49, 53, 57, 61.

120. How many terms are in the arithmetic sequence above?

120. _____

121. What is the median of the sequence above?

121. _____

122. What is the sum of the terms in the sequence above?

122. _____

123. How could you find the average of the terms in the arithmetic sequence above using only the first term and the last term? Explain.

PRACTICE | Answer each question below.

124. The middle term of a 25-term arithmetic sequence is 43. What is the sum of the first and last terms of the sequence?

124. _____

125. What is the sum of the terms in a 30-term arithmetic sequence with first term 19 and last term 81?

125. _____

126. Find the sum of all 100 terms in an arithmetic sequence that has first term 24 and common difference 4.

126. _____

127. What is the sum of the arithmetic series $15+21+27+\cdots+207$?

127. _____

EXAMPLE | Grogg makes a pattern out of gumballs, as shown below. How many gumballs are needed to make the 33rd figure in his pattern?

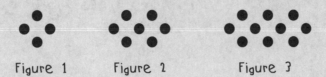

Figure 1   Figure 2   Figure 3

The first figure in Grogg's pattern has 4 gumballs. Then, each figure has 3 more gumballs than the figure that came before.

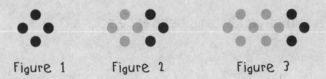

Figure 1   Figure 2   Figure 3

The number of gumballs in each figure of Grogg's pattern makes an arithmetic sequence with first term 4 and common difference 3.

So, Grogg's 33rd figure will have $4+32(3) = 4+96 = $ **100** gumballs.

PRACTICE | Answer the following questions about the gumball pattern below.

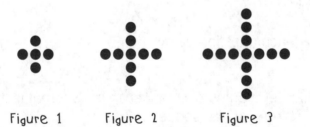

Figure 1   Figure 2   Figure 3

128. How many gumballs are needed to make the 8th figure in this pattern?

128. _____

129. Write a simplified expression for the number of gumballs needed to make the nth figure in this pattern.

129. _____

130. There are 221 gumballs in the kth figure of this pattern. What is k?

130. $k = $ _____

PRACTICE | Answer each question about gumball patterns below.

131. The first four figures in a gumball pattern are shown below. How many total gumballs are needed to make the 30th figure of this pattern?

131. _____

Figure 1   Figure 2   Figure 3   Figure 4

132. How many gumballs are needed to make the 10th figure in the pattern shown below?
★

132. _____

Figure 1   Figure 2   Figure 3   Figure 4

133. The kth figure of the pattern below uses 420 gumballs. What is k?
★

133. $k =$ _____

Figure 1   Figure 2   Figure 3   Figure 4

134. In the gumball pattern below, some gumballs are black and some gumballs are gray. Write an expression for the **total** number of gumballs in the nth figure of this pattern.
★

134. _____

Figure 1   Figure 2   Figure 3   Figure 4

In the **Fibonacci sequence**, the first term is 1 and the second term is 1. After that, every term is the sum of the two previous terms.

EXAMPLE | List the first five terms of the Fibonacci sequence.

The first two terms of the Fibonacci sequence are 1 and 1. To find the 3rd term in the sequence, we add the 1st and 2nd terms: $1+1=2$.

$$\underline{1}, \underline{1}, 2, ...$$

Then, the 4th term is the sum of the 2nd and 3rd terms: $1+2=3$.

$$1, \underline{1}, \underline{2}, 3, ...$$

Finally, the 5th term is the sum of the 3rd and 4th terms: $2+3=5$. So, the first five terms of the Fibonacci sequence are

$$\textbf{1, 1, 2, 3, 5, ...}$$

Squawk. Fib-uh-nah-chee.

PRACTICE | Answer each question below.

135. Fill in the blanks below to list the first 10 terms of the Fibonacci sequence.

$$1, 1, 2, 3, 5, \underline{\hspace{1cm}}, \underline{\hspace{1cm}}, \underline{\hspace{1cm}}, \underline{\hspace{1cm}}, \underline{\hspace{1cm}}$$

136. The 19th and 20th terms of the Fibonacci sequence are 4,181 and 6,765. What is the 18th term of the Fibonacci sequence?

136. _____

137. Is the 100th term of the Fibonacci sequence odd or even? Explain.

138. Grogg writes a sequence of terms similar to the Fibonacci sequence, in which each term is the sum of the two terms before it. However, instead of starting with 1, the first term in Grogg's sequence is 18. If the fourth term in Grogg's sequence is 76, what is Grogg's second term?

138. _____

PRACTICE | Answer each question below about Hoppy the coilbeast.

Hoppy the coilbeast is hopping up a flight of
stairs. He can hop up 1 or 2 steps at a time.

139. Hoppy can reach the 1st step in 1 way: by taking a 1-step hop from the base.
Hoppy can reach the 2nd step in 2 ways: by taking a 2-step hop from the
base, or by taking two 1-step hops from the base.
How many ways can Hoppy go from the base of the steps to the 3rd step?

139. _____

140. How many different ways are there for Hoppy to go from the base of
the steps to the 4th step?

140. _____

141. To reach the 5th step, Hoppy can hop to the 4th step and take a 1-step
★ hop from there, or he can hop to the 3rd step and take a 2-step hop.
How many ways can Hoppy go from the base of the steps to the 5th step?

141. _____

142. How many ways are there for Hoppy to get to the 6th, 7th, 8th, 9th, and 10th steps?
★ The first two entries have been filled for you, and you can use your answers above
for the 3rd, 4th, and 5th steps.

$$\frac{1}{1^{st}}, \frac{2}{2^{nd}}, \frac{}{3^{rd}}, \frac{}{4^{th}}, \frac{}{5^{th}}, \frac{}{6^{th}}, \frac{}{7^{th}}, \frac{}{8^{th}}, \frac{}{9^{th}}, \frac{}{10^{th}}, \dots$$

143. How many ways are there to arrange 10 dominos to form a 2-by-10
★ rectangle of dominos? Two possible arrangements are shown below.
★

143. _____

PRACTICE | For the problems below, use the given rule to fill in the blanks of each sequence.

144. Each term is half the product of the two previous terms.

8, 1, _____, _____, _____, _____, _____, _____

145. Each term is the opposite of twice the previous term.

5, _____, _____, _____, _____, _____ , _____

146. Each term is the reciprocal of the sum of the previous two terms.

0, 1, _____, _____, _____, _____, _____

147. Each term is the square of the sum of the previous term's digits.

65, _____, _____, _____, _____, _____, _____ , _____

148. Each term is the sum of the squares of the previous term's digits.

65, _____, _____, _____, _____, _____, _____ , _____

149. Leodore writes a sequence whose first two terms are 0 and 1. After
★ that, each term is the sum of all previous terms. Write an expression
that correctly gives the nth term of Leodore's sequence for all values
of n that are greater than 2.

149. _____

PRACTICE | Use the rules given below to answer the questions that follow.

To find the next term in a Lizzie sequence, reverse the digits of the current term and add 1 to the result. For example, the term that comes after 52 is $25 + 1 = 26$.

150. Fill in the missing terms in each Lizzie sequence below.

    a.   32, _____, _____, _____, _____, _____, _____, _____, _____, _____

    b.   _____, _____, _____, _____, _____, _____, _____, _____, _____, 100

To find the next term in an Alex sequence, take the two previous terms and subtract the smaller one from the larger one (order does not matter if the terms are equal). For example, if the first two terms are 45 and 60, the third term is $60 - 45 = 15$.

151. Fill in the missing terms in each Alex sequence below.

    a.   85, 52, _____, _____, _____, _____, _____, _____, _____, _____

    b.   76, 47, _____, _____, _____, _____, _____, _____, _____, _____

152. ★ What is the 75ᵗʰ term of an Alex sequence whose first two terms are 5 and 3?

152. _____

153. ★ Fill in the blanks below to create a 10-term Alex sequence in which every term is smaller than the one before it.

    _____, _____, _____, _____, _____, _____, _____, _____, _____, _____

154. ★ Ralph makes a sequence that begins with a 2-digit number. After the first term, each term is 9 times the sum of the digits of the term before it. How many different numbers could be the 4ᵗʰ term in Ralph's sequence?

154. _____

Look for patterns that will help you solve these problems.

PRACTICE | Answer each question below.

155. What is the units digit of 3^{99}?

155. _____

156. Compute the sum of the powers of 2 listed below:

$1+2+4+8+16+32+64+128+256+512+1{,}024+2{,}048+4{,}096$

156. _____

157. If the pattern continues below, under which letter will the number 99 appear?

157. _____

A	B	C	D	E	F
1	2	3	4	5	6
12	11	10	9	8	7
13	14	15	16	17	18
24	23	22	21	20	19
⋮	⋮	⋮	⋮	⋮	⋮

158. ★ If the pattern continues below, what number will be directly above the number 100?

158. _____

```
21—22—23—24— ···
|
20   7— 8— 9—10
|    |          |
19   6    1— 2  11
|    |       |   |
18   5— 4— 3   12
|               |
17—16—15—14—13
```

PRACTICE | Answer each question below.

159. What is the remainder when 2^{100} is divided by 7?

159. _____

160. A compubot has a button marked ★. When an odd integer is
★   displayed, pressing ★ multiplies the number by 3 and adds 1. When
    an even integer is displayed, pressing ★ divides the number by 2. If
    the compubot displays 3 now, what will it display after the ★ button is
    pressed 75 times?

160. _____

161. Grogg has two 2-cup juice boxes. The first box is full and the second is
★   empty. Grogg pours half of the juice in the first box into the second.
    Then, he pours one third of the juice in the second box into the first.
    Then, he pours one fourth of the juice in the first box into the second.
    Then, he pours one fifth of the juice in the second box into the first.
    If he continues to pour juice from one juice box to the other this way, how
    many cups of juice will be in the first juice box after Grogg's 99th pour?

161. _____

PRACTICE | Answer each question below.

162. What is the smallest positive number that appears in the infinite arithmetic sequence that has first term -999 and common difference 8?

162. _____

$$-999, -991, -983, -975, -967, \ldots$$

163. Write the next five numbers in the following sequence:

$$1, \ 10, \ 11, \ 100, \ 101, \ 110, \ 111, \ \rule{1cm}{0.4pt}, \ \rule{1cm}{0.4pt}, \ \rule{1cm}{0.4pt}, \ \rule{1cm}{0.4pt}, \ \rule{1cm}{0.4pt}, \ \ldots$$

164. ★ The sequence below is the list of positive integers that are not perfect squares. What is the 50th term in this sequence?

164. _____

$$2, 3, 5, 6, 7, 8, 10, \ldots$$

165. ★ The sequence below is made by adding 1, then 2, then 3, then repeating this over and over. What is the smallest 7-digit number that appears in this sequence?

165. _____

$$0, 1, 3, 6, 7, 9, 12, 13, 15, 18, \ldots$$

PRACTICE | Answer each question below.

166. What is the greatest possible term that can appear in an arithmetic
★ sequence of 7 positive integers whose sum is 133?

166. _____

167. In a ***Tribonacci sequence***, each term is the sum of the ***three***
★ previous terms. Consecutive terms in a Tribonacci sequence are
shown below. What is x?

$$x, y, z, 44, 81, \ldots$$

167. $x =$ _____

168. The first three figures in a toothpick pattern are shown below. How
★ many toothpicks are needed to make the 20th figure in this pattern?

168. _____

169. Bronkle writes the arithmetic sequence 40, 46, 52, ...
★ Gergum writes the arithmetic sequence -50, -41, -32, ...
The kth term of Bronkle's sequence is equal to the kth term of
Gergum's sequence. What is the kth term of both sequences?

169. _____

CHAPTER 8
Ratios & Rates

Use this Practice book with
Guide 5C from BeastAcademy.com.

Recommended Sequence:

Book	Pages:
Guide:	42-51
Practice:	37-47
Guide:	52-57
Practice:	48-57
Guide:	58-63
Practice:	58-63
Guide:	64-69
Practice:	64-73

You may also read the entire chapter
in the Guide before beginning the
Practice chapter.

Ratios are used to show how quantities relate.

For example, if a bowl of fruit contains 5 apples and 7 bananas, we say that the ratio of apples to bananas in the bowl is 5 to 7.

Similarly, we could say that the ratio of bananas to apples is 7 to 5.

Pay attention to the order of the terms in a ratio.

PRACTICE | Answer each question about the collection of shapes below.

1. What is the ratio of black circles to white circles?

1. _____ to _____

2. What is the ratio of circles to triangles?

2. _____ to _____

3. What is the ratio of black shapes to white shapes?

3. _____ to _____

4. What is the ratio of white triangles to black triangles?

4. _____ to _____

5. How many of the circles below must be shaded so that there are 2 shaded circles for every 1 unshaded circle?

5. _____

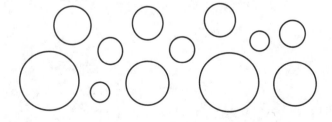

Ratios

Ratios are usually written with the quantities separated by a colon. For example, a 2-to-3 ratio of apples to oranges is usually written 2:3.

Ratios are almost always written in **simplest form**, meaning the quantities in the ratio are integers that do not have any common factors greater than 1.

| **EXAMPLE** | Mr. Underhill has 6 cats and 8 dogs. What is the ratio of cats to dogs, in simplest form? |

The ratio of cats to dogs is 6:8. However, since 6 and 8 have a common factor of 2, this ratio is not in simplest form.

We can split the animals into two groups, each with $6 \div 2 = 3$ cats and $8 \div 2 = 4$ dogs. So, there are 3 cats for every 4 dogs.

The greatest common factor of 3 and 4 is 1, so we cannot simplify any further. The ratio of cats to dogs is **3:4**.

We can write 6:8 = 3:4.

| **PRACTICE** | Express each ratio below in simplest form. |

6. 5:10 = _____

7. 197:197 = _____

8. 25:9 = _____

9. 40:16 = _____

10. 42:70 = _____

11. 121:88 = _____

12. 24:92 = _____

13. 91:65 = _____

14. ★ $2\frac{1}{3} : 7 =$ _____

15. ★ $\frac{1}{3} : \frac{3}{5} =$ _____

PRACTICE | Answer each question below. Write all ratios in simplest form.

16. Ted has 16 blue toy cars and 10 red toy cars. What is the ratio of blue cars to red cars?

16. _____

17. In a class of 30 students, 26 of the students have brown eyes, and the rest have green eyes. What is the ratio of brown-eyed students to green-eyed students?

17. _____

18. In 100 flips of a coin, 56 of the flips land heads. What is the ratio of heads flipped to tails flipped?

18. _____

19. The ratio of the side length of a small square to the side length of a larger square is 3:5. What is the ratio of the area of the small square to the area of the large square?

19. _____

20. Raquel makes a smoothie with $\frac{2}{3}$ of a cup of yogurt and $\frac{1}{2}$ of a cup of strawberries. What is the ratio of cups of yogurt to cups of strawberries?

20. _____

| EXAMPLE | The ratio of chickens to cows on a farm is 2 to 3. If there are 54 chickens, how many cows are there? |

The ratio tells us that for every 2 chickens on the farm, there are 3 cows. So, we can split the chickens and cows into groups, each with 2 chickens and 3 cows.

Since there are 54 chickens, we can make $54 \div 2 = 27$ groups with 2 chickens each. Each of the 27 groups also has 3 cows.
So, there are $27 \cdot 3 = \textbf{81}$ cows.

All together, the farm has $27 \cdot \underline{2} = 54$ chickens and $27 \cdot \underline{3} = 81$ cows.

| PRACTICE | Answer each question below. |

21. The ratio of dragons to yetis in a dance class is 5 to 3. If there are 30 dragons, how many yetis are there?

21. _____

22. At a math tournament, the ratio of Beast Academy fans to Orb Academy fans is 5 to 4. If everyone at the tournament is a fan of just one school and there are 160 Beast Academy fans, how many Orb Academy fans are there?

22. _____

23. The ratio of grape sodas to orange sodas sold from a soda machine last week was 7 to 6. If 84 orange sodas were sold, how many grape sodas were sold?

23. _____

PRACTICE | Answer each question below.

24. On Priti's last history test, she answered 2 questions incorrectly for every 5 questions that she answered correctly. If Priti had 30 correct answers, how many total questions were on the test?

24. _____

25. Roland baked sugar cookies and oatmeal-raisin cookies in a 4-to-3 ratio. If Roland baked 36 oatmeal-raisin cookies, how many cookies did he bake all together?

25. _____

26. ★ Allison makes green paint by mixing blue paint with yellow paint in a 2:3 ratio. How many pints of blue paint must Allison mix with $\frac{1}{4}$ pint of yellow paint to make green paint?

26. _____

27. ★ On a recent trip, Peter traveled 7 miles by train for every 3 miles he traveled by bus. He also biked 1 mile for every 14 miles he took a bus or train. If he traveled 210 miles by bus, how far did he bike?

27. _____

28. ★ Anna has 40 white beads and 90 black beads. She wants to make a necklace in which the ratio of white to black beads is 3 to 7. What is the greatest number of white beads she can use to make the necklace?

28. _____

There are many ways to reason through problems involving ratios.

In many situations, it helps to consider how the parts in a ratio relate to the **whole** amount.

EXAMPLE The ratio of soda to juice in Daryl's fruit punch recipe is 3 to 7. How much soda and juice does he need to make 50 fluid ounces of fruit punch?

For every 3 fluid ounces of soda in the fruit punch, there are 7 fluid ounces of juice. So, in every 10 fluid ounces of punch, 3 fluid ounces are soda and 7 fluid ounces are juice.

A 50-fl-oz pitcher holds $50 \div 10 = 5$ servings of 10 fluid ounces, each with 3 fluid ounces of soda and 7 fluid ounces of juice. So, Daryl needs $\underline{3} \cdot 5 = $ **15 fl oz of soda** and $\underline{7} \cdot 5 = $ **35 fl oz of juice**.

— *or* —

For every 3 parts soda, there are 7 parts juice, for a total of $3 + 7 = 10$ parts.

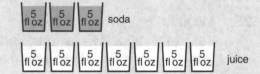

If we divide 50 fluid ounces of punch into 10 equal parts, each part is 5 fluid ounces. So, the 3 parts of soda equal $\underline{3} \cdot 5 = $ **15 fl oz** of the punch, and the 7 parts juice equal $\underline{7} \cdot 5 = $ **35 fl oz**.

— *or* —

In a punch that is 3 parts soda and 7 parts juice, 3 of the 10 parts are soda and 7 of the 10 parts are juice. In other words, $\frac{3}{10}$ of the punch is soda, and $\frac{7}{10}$ of the punch is juice. So, to make 50 fluid ounces of punch, Daryl needs $\frac{3}{10} \cdot 50 = $ **15 fl oz of soda** and $\frac{7}{10} \cdot 50 = $ **35 fl oz of juice**.

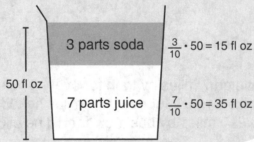

PRACTICE Answer the question below.

29. Every blorble has 3 long horns and 4 short horns. There are 98 horns on all of the blorbles in a field. How many of those horns are short?

29. _____

PRACTICE | Answer each question below.

30. At a basketball game, every player is wearing a white or blue jersey. The ratio of white jerseys to blue jerseys is 4 to 5. How many of the 27 players at the game are wearing white jerseys?

30. _____

31. In a junior Beastball league, each team has 3 defenders and 5 strikers. How many of the 96 players in the league are defenders?

31. _____

32. Edna bakes 240 cupcakes and arranges them onto trays so that each tray holds 4 chocolate and 6 vanilla cupcakes. How many vanilla cupcakes did she make?

32. _____

33. ★ A zoo has an exhibit with spotted stingravens, striped stingravens, and pandakeets. The ratio of spotted stingravens to striped stingravens is 3:5, and the ratio of stingravens to pandakeets is 2:9. If there are 176 animals in the exhibit, how many are striped stingravens?

33. _____

34. ★ A rectangle has a width-to-height ratio of 5:2. The perimeter of the rectangle is 210 inches. What is the area of the rectangle, in square inches?

34. _____

EXAMPLE | Which ratio below is equivalent to 9:36?

4:18, 22:88, 24:94, or 21:98.

Equivalent ratios have the same simplest form.

In simplest form, we have

4:18 = 2:9, 22:88 = 1:4, 24:94 = 12:47, and 21:98 = 3:14.

Since 9:36 = 1:4 and 22:88 = 1:4, we have 9:36 = 22:88.

So, **22:88** is equivalent to 9:36.

PRACTICE | Answer each question below.

35. Circle the ratio below that is not equivalent to the other four.

21:28   63:84   39:52   24:36   51:68

36. Circle the ratios below that are equivalent to 11:17.

22:34   30:48   33:51   44:68   55:85   55:88   60:93   66:102

37. Use the 8 ratios below to write four pairs of equivalent ratios.

21:35   14:24   32:48   56:96

55:90   33:54   12:20   18:27

37. _____ = _____

_____ = _____

_____ = _____

_____ = _____

38. Use the numbers 6, 9, 28, 30, 42, and 140 to write three equivalent ratios.

_____ : _____   =   _____ : _____   =   _____ : _____

PRACTICE | Answer each question below.

39. Fill the missing entries in the chart below so that the ratio of milk to butter in each column is the same.

ounces of milk			27	36	54	
ounces of butter	14		56		126

40. There are 18 pigs and 24 goats in a pen. Nine pigs and some goats enter the pen, but the ratio of pigs to goats does not change. How many goats entered the pen?

40. _____

41. Captain Kraken has 36 gold coins and 63 silver coins in a satchel. After spending some of the coins, Captain Kraken sees that the ratio of gold coins to silver coins has not changed. If Kraken spent 8 gold coins, how many silver coins did he spend?

41. _____

42. ★ Grogg mixes $\frac{3}{4}$ of a cup of cornstarch with $\frac{1}{2}$ of a cup of water to make $1\frac{1}{4}$ cups of oobleck. If he adds 1 cup of cornstarch to the mixture, how many cups of water should he add to maintain the ratio of cornstarch to water in his oobleck?

42. _____

In a **Rectivide** puzzle, the goal is to divide a single rectangle into *three* smaller rectangles so that each small rectangle has the same ratio of gray squares to white squares as the original rectangle.

EXAMPLE | Solve the Rectivide puzzle below.

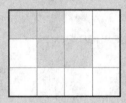

The ratio of gray squares to white squares in the original rectangle is 4:8 = 1:2. So, we can make groups of 3 squares in each of the smaller rectangles, with 1 gray and 2 white squares. Therefore, the area of each small rectangle is a multiple of 3.

We can split the rectangle into three smaller rectangles as shown below so that the ratio of gray squares to white squares is 1:2 in each smaller rectangle.

This is the
only solution.

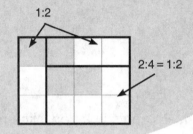

1:2

2:4 = 1:2

PRACTICE | Solve each Rectivide puzzle below.

43.

44.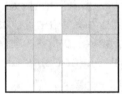

45.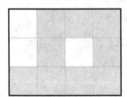

46.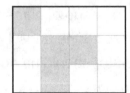

PRACTICE | Solve each Rectivide puzzle below.

47.

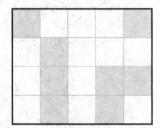

48.

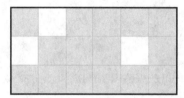

49.

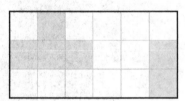

50.

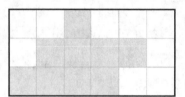

51.

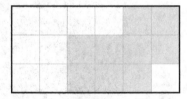

52.

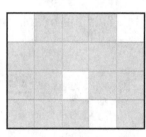

53.

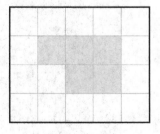

54.

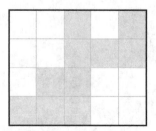

We often use fractions when working with ratios. For example, a boy:girl ratio of 2:3 can be written as

$$\frac{\text{boys}}{\text{girls}} = \frac{2}{3}.$$

This means that the number of boys divided by the number of girls equals $\frac{2}{3}$.

A **proportion** is an equation showing that two ratios are equal. There are many ways to find the missing value in a proportion.

To solve for x in the proportion $4:7 = x:21$, we can solve for x in the equation $\frac{4}{7} = \frac{x}{21}$.

EXAMPLE | What is the value of x in the equation below?

$$\frac{4}{7} = \frac{x}{21}$$

We can convert the fraction.

We can write $\frac{4}{7}$ with a denominator of 21 by multiplying the numerator and denominator by 3.

$\frac{4}{7} = \frac{12}{21}$, so $x = \mathbf{12}$.

$$\frac{4}{7} \overset{\cdot 3}{\underset{\cdot 3}{=}} \frac{x}{21}$$

— *or* —

We can isolate the variable.

To isolate the variable x, we multiply both sides of the equation by 21.

On the left side, we get $4 \cdot 3 = 12$. So, $x = \mathbf{12}$.

$$\frac{4}{7} \cdot 21 = \frac{x}{21} \cdot 21$$

$$\frac{4}{7} \cdot 21^{3} = \frac{x}{\cancel{21}} \cdot \cancel{21}$$

$$12 = x$$

PRACTICE | Fill in the missing value in each equation below.

55. $\dfrac{2}{3} = \dfrac{}{15}$

56. $\dfrac{15}{} = \dfrac{5}{12}$

57. $\dfrac{}{32} = \dfrac{7}{8}$

58. $\dfrac{19}{11} = \dfrac{}{88}$

PRACTICE | Solve for the variable in each equation below.

59. $\dfrac{3}{7} = \dfrac{x}{42}$

60. $\dfrac{13}{9} = \dfrac{a}{45}$

61. $\dfrac{w}{35} = \dfrac{21}{15}$

62. $\dfrac{m}{21} = \dfrac{9}{4}$

59. $x =$ _____

60. $a =$ _____

61. $w =$ _____

62. $m =$ _____

EXAMPLE | What is the value of x in the equation below?

$$\frac{5}{9} = \frac{8}{x}$$

We can eliminate the denominators.

We eliminate the denominators of $\frac{5}{9}$ and $\frac{8}{x}$ by multiplying both sides of the equation by a common multiple of their denominators: $9x$.

This gives $5x = 72$.

We divide both sides by 5 to get $x = \frac{72}{5} = 14\frac{2}{5}$.

$$\frac{5}{\cancel{9}} \cdot \cancel{9}x = \frac{8}{\cancel{x}} \cdot 9\cancel{x}$$

$$5 \cdot x = 8 \cdot 9$$

$$5x = 72$$

$$x = \frac{72}{5}$$

$$x = 14\frac{2}{5}$$

For any equation $\frac{a}{b} = \frac{c}{d}$, we have $ad = bc$.

PRACTICE | Solve for the variable in each equation below. Write your answer in simplest form.

63. $\dfrac{2}{3} = \dfrac{15}{m}$

64. $\dfrac{12}{s} = \dfrac{5}{8}$

63. $m =$ _____

64. $s =$ _____

65. $\dfrac{7}{4} = \dfrac{15}{c}$

66. $\dfrac{14}{a} = \dfrac{4}{9}$

65. $c =$ _____

66. $a =$ _____

67. $\dfrac{2}{5} = \dfrac{15}{z}$

68. $\dfrac{3}{10} = \dfrac{10}{v}$

67. $z =$ _____

68. $v =$ _____

69. $\dfrac{8}{n} = \dfrac{11}{6}$

70. $\dfrac{10}{7} = \dfrac{6}{r}$

69. $n =$ _____

70. $r =$ _____

EXAMPLE

Katie makes purple paint by mixing blue paint and red paint in a 3:5 ratio. How many ounces of red paint should she mix with 8 ounces of blue paint to make purple paint?

The ratio of blue paint to red paint is 3:5.

The ratio of 8 ounces of blue paint to ounces of red paint (r) is 8:r.

Since we want 3 parts blue paint for every 5 parts red paint, we have 8:r = 3:5.

We express 8:r = 3:5 with fractions.

Multiplying both sides of $\frac{8}{r} = \frac{3}{5}$ by $5r$ gives $40 = 3r$.

Then, dividing both sides of the equation by 3 gives $\frac{40}{3} = r$. So, $r = \frac{40}{3} = 13\frac{1}{3}$.

So, Katie should mix **$13\frac{1}{3}$** ounces of red paint with 8 ounces of blue paint.

$$\frac{8}{r} = \frac{3}{5} \leftarrow \text{blue paint} \atop \leftarrow \text{red paint}$$

$$\frac{8}{r} \cdot 5r = \frac{3}{5} \cdot 5r$$

$$40 = 3r$$

$$\frac{40}{3} = r$$

$$13\frac{1}{3} = r$$

Many problems involving ratios can be solved by writing an equation and solving for an unknown value!

— *or* —

We can instead use the ratios of red paint to blue paint: r:8 = 5:3. Then, we have

$$\frac{r}{8} = \frac{5}{3} \leftarrow \text{red paint} \atop \leftarrow \text{blue paint}$$

Multiplying both sides by 8 gives $r = \frac{40}{3} = 13\frac{1}{3}$.

So, Katie should mix **$13\frac{1}{3}$** ounces of red paint with 8 ounces of blue paint.

PRACTICE

Solve each problem below. Express all non-integer answers as mixed numbers in simplest form.

71. Kristen writes a sentence that has 5 consonants for every 4 vowels. If her sentence has 40 consonants, how many vowels does it have?

71. _____

72. To make fudge, Cori uses 9 cups of sugar for every 2 cups of cocoa. If she uses 5 cups of sugar, how many cups of cocoa should she use?

72. _____

73. The ratio of the width of a Blorgbeast's hoof to the height of the Blorgbeast's hoof is 3:8. How many inches wide is a Blorgbeast hoof that is 5 inches tall?

73. _____

PRACTICE | Solve each problem below. Express all non-integer answers as mixed numbers in simplest form.

74. Hummingbeast nectar is mixed in a ratio of 2 parts sugar for every 7 parts water. How many ounces of sugar should be added to 16 ounces of water to make hummingbeast nectar?

74. _____

75. Terry hits a home run 2 out of every 7 times that he hits a baseball. If he hit 28 home runs this season, how many times did he hit the ball but not get a home run?

75. _____

76. The ratio of the area of isosceles right triangle ABC to the area of right triangle DEF is 3:5. How many inches long is side ED?

76. _____

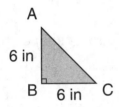

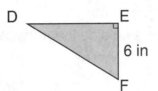

77. The ratio of the weight of a biffo to the weight of a triffo is 2:3. If a group of 30 biffos weighs 40 pounds, how many pounds does a group of 30 triffos weigh?

77. _____

78. ★ Will writes two numbers, a and b. The ratio of a to b is 3 to 5. The sum of a and b is 100. What are a and b?

78. $a =$ _____

$b =$ _____

79. ★ The ratio of goals scored by Ben to goals scored by Alfie last season was 5:4. If Ben scored 16 more goals than Alfie last season, how many goals did Alfie score?

79. _____

Two shapes are *similar* if one shape can be flipped, rotated, or shrunk to exactly match the other.

Similar shapes have sides that are *proportional*. This means that the ratios of the lengths of the corresponding sides are the same.

EXAMPLE | The rectangles below are similar. Find the missing side length.

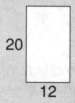

The ratio of the short side of the first rectangle to the short side of the second rectangle is $h{:}12$.

The ratio of the long side of the first rectangle to the long side of the second rectangle is $25{:}20$.

The rectangles are similar, so $h{:}12 = 25{:}20$. This means $\frac{h}{12} = \frac{25}{20}$. Solving for h, we have $h = \mathbf{15}$.

PRACTICE | Find the missing side length(s) in each pair of similar shapes below.

80.

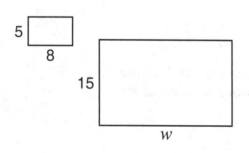

80. $w =$ _____

81.

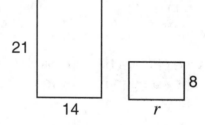

81. $r =$ _____

82.

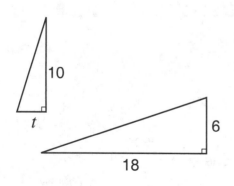

82. $t =$ _____

83.

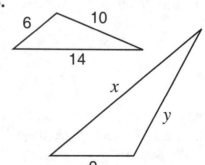

83. $x =$ _____

$y =$ _____

PRACTICE | Answer the questions below.

84. The ratio of the edge length of a small cube to the edge length of a large cube is 5:7. What is the ratio of the surface area of the small cube to the large cube?

84. _____

85. Chris has a model train set in which all the pieces are built to scale. This means that the ratios of the sizes of his models to the sizes of the actual objects are always the same. His $3\frac{3}{4}$-inch-tall station models a real station that is 50 feet tall.

a. One of Chris's model train cars is 6 inches long. How many feet long is the actual train car?

a. _____

b. How many feet long is Chris's model of an actual track that is 12,000 feet long?

b. _____

86. The map below models the distance between four cities around Route 43 and Route 2. Yetiville and North Pegasus are 25 miles apart, which is represented on the map by a distance of 3 inches.

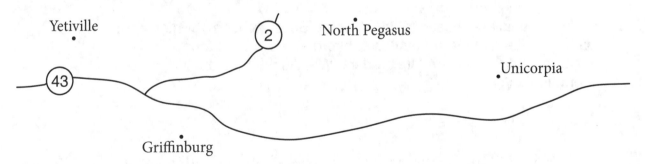

a. ★ Griffinburg and North Pegasus are $2\frac{1}{4}$ inches apart on this map. How many miles apart are these two cities?

a. _____

b. ★ The town of West Mermaid is 40 miles from Unicorpia. If West Mermaid were included on this map, how many inches away from Unicorpia would it be?

b. _____

EXAMPLE | A bouquet of flowers contains 4 roses, 8 daffodils, and 14 sunflowers. What is the ratio of roses to daffodils to sunflowers?

The ratio of roses to daffodils to sunflowers is 4 to 8 to 14 or 4:8:14. Since 4, 8, and 14 all have a common factor of 2, we simplify this ratio:

$$4:8:14 = \mathbf{2:4:7}.$$

So, we can make groups of 13 flowers, each with 2 roses, 4 daffodils, and 7 sunflowers.

*Remember that a ratio is in simplest form when the greatest common factor of **all** terms in the ratio is 1.*

We can write ratios to describe the relationships between **more than** two quantities.

PRACTICE | Solve each problem below. Write all ratios in simplest form.

87. In a room of 93 students, 18 are wearing red shirts, 30 are wearing blue shirts, and 45 are wearing yellow shirts. What is the ratio of red to blue to yellow shirts?

87. _____

88. Sixty students are chosen to march in the Beast Academy winter parade. Fifteen of the students are third graders, 14 are fourth graders, and the rest are fifth graders. What is the ratio of third to fourth to fifth graders?

88. _____

89. Ernie has three types of fish in his tank: guppies, goldfish, and angelfish. The ratio of guppies to goldfish to angelfish in the tank is 8 to 5 to 3. What fraction of the fish in Ernie's tank are goldfish?

89. _____

90. ★ Rolfe grows peaches, Fuji apples, and Gala apples in his orchard. The ratio of Fuji to Gala trees is 3:5, and the ratio of apple to peach trees is 4:3. What is the ratio of Fuji to Gala to peach trees?

90. _____

PRACTICE | Solve each problem below. Write all ratios in simplest form.

91. A recipe for 36 cookies calls for $1\frac{1}{2}$ cups of butter, $2\frac{1}{2}$ cups of sugar, and $2\frac{2}{3}$ cups of flour. What is the ratio of butter to sugar to flour?

91. _____

92. The ratio of length to width to height of a rectangular prism is 5:6:8. If the prism is 60 cm wide, what is the surface area of the prism, in square centimeters?

92. _____

93. The ratio of red to blue to yellow marbles in a bag is 1:2:3. If there are 16 blue marbles, how many marbles are in the bag?

93. _____

94. ★ At Hoppy Pond, the ratio of frogs to toads is 3:5, and the ratio of green frogs to yellow frogs is 3:2. If every frog is either green or yellow, what is the ratio of green frogs to yellow frogs to toads?

94. _____

95. ★ The ratio of a to b to c is 2:9:3. If the sum $a+b+c$ is 70, what is the product abc?

95. _____

96. ★★ The ratio of length to width to height of a rectangular prism is 2:5:8. If the volume of the prism is 2,160 cubic inches, what are the length, width, and height of the prism?

96. Length: _____

Width: _____

Height: _____

In a **Ratiotile** puzzle, the goal is to fill every square and triangle so that
- each triangle contains a positive integer, and
- each square gives the ratio of the smaller number to the larger number in the adjacent triangles, in simplest form.

EXAMPLE | Complete the Ratiotile puzzle to the right.

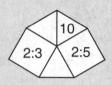

The ratio of the triangles that touch the right square is 2:5. If 10 is the larger of the two numbers, we have $\boxed{4}$:10 = 2:5. If 10 is the smaller of the two numbers, we have 10:$\boxed{25}$ = 2:5. So, the value in the center triangle is 4 or 25.

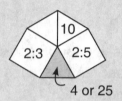

The ratio of the numbers in the triangles that touch the left square is 2:3. Since 25 is not a multiple of 2 or 3, there is no integer that has a 2:3 ratio with 25. So, the center triangle is 4.

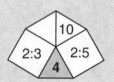

No integer smaller than 4 has a 2:3 ratio with 4. However, if 4 is the smaller number in the 2:3 ratio, we have 4:$\boxed{6}$ = 2:3. So, the top-left triangle is 6.

We check that 4:6 = 2:3 and 4:10 = 2:5. ✓

PRACTICE | Solve each Ratiotile puzzle below.

97.

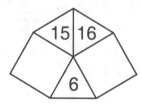

98.

99.

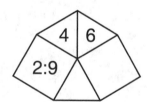

100.

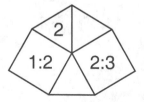

101.

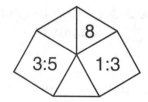

102.

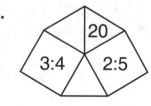

Remember that each ratio describes the relationship of the smaller number to the larger number!

PRACTICE | Solve each Ratiotile puzzle below.

103.

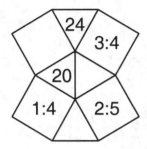

104.

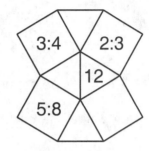

105.

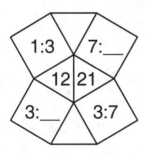

106.

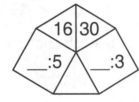

107.

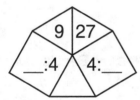

108.

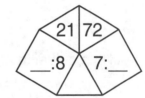

PRACTICE | The squares below give the ratio of the numbers in the three adjacent triangles in order from smallest to largest.

109.
★

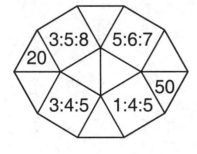

110.
★

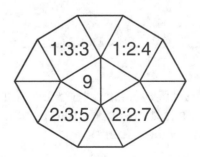

A **rate** describes how much of one quantity there is for one unit of another quantity.

We often use the word "per" in rates. "Per" means "for every" or "for each."

The following are all examples of rates:

65 miles per hour   5 dollars per gallon   90 sheets per roll

19 points per game   57 words per minute   21 students per class

10 dollars per week   32 miles per gallon   12 markers per box

EXAMPLE | Yergel scored 136 points in 8 games.
How many points per game is this?

$$\frac{136 \text{ points}}{8 \text{ games}} = \frac{136}{8} \text{ points per game} = \textbf{17 points per game}.$$

We may also write this as **17 points/game** or $17 \frac{\textbf{points}}{\textbf{game}}$.

PRACTICE | Answer each question below.

111. Beasty brand paper towels have 85 sheets per roll. How many sheets are in...

2 rolls? _____   5 rolls? _____   10 rolls? _____   55 rolls? _____

112. Bess's car uses 5 gallons of fuel to drive 160 miles. How many miles will Bess's car go using...

1 gallon? _____   3 gallons? _____   10 gallons? _____   55 gallons? _____

113. Grogg buys 6 pens at the school store for $2.10. At this rate, what is the cost of...

1 pen? _____   5 pens? _____   10 pens? _____   55 pens? _____

114. Arthur's Grocery sells a 16-pound bag of potatoes for $12. Betty's Grocery sells a 10-pound bag of potatoes for $8. Which store's potatoes cost less per pound?

114. _____

PRACTICE | Answer each question below.

115. Benji types his 1,500-word book report in 30 minutes. At this rate, how many minutes will it take for him to type his 5,000-word history essay?

115. _____

116. Forty-five baby dragons were born in the last 9 days on Beast Island. At this rate, how many baby dragons will be born in the next 12 days on Beast Island?

116. _____

117. Freddy is baking pies for a bake sale. He buys a 15-pound bag of 48 apples for $18. He uses all of the apples and 24 ounces of sugar to bake 8 pies. Write each rate below.

a. _____ apples per pie

b. _____ ounces of sugar per pie

c. _____ dollars per pound of apples

d. _____ pounds of apples per pie

e. _____ ounces of sugar per apple

f. _____ pounds per apple

Use the following information for problems 118-121:

Mr. Jones exchanges 75 Beastbucks for 300 Torts, the currency of Tortuga Island.

118. How many Torts does he receive per Beastbuck?

118. _____ Torts per Beastbuck

119. How many Torts will Mr. Jones receive in exchange for 100 Beastbucks?

119. _____

120. At this rate, when Mr. Jones exchanges Beastbucks for Torts, how many Beastbucks will he receive per Tort?

120. _____ Beastbucks per Tort

121. How many Beastbucks will Mr. Jones receive in exchange for 480 Torts?

121. _____

Speed is a type of rate that describes a relationship between distance and time.

We often express speed as the distance traveled for one unit of time. For example, one common unit of speed is miles per hour (mph), which describes how many miles are traveled in one hour.

EXAMPLE | Fiona can sprint 200 meters in 25 seconds.
What is Fiona's speed in meters per second?

Fiona's speed is the ratio of the distance she ran to the amount of time it took her to run that distance.

Since Fiona can run 200 meters in 25 seconds, she can run $200 \div 25 = 8$ meters in one second. So, her speed is **8 meters per second**.

$$\begin{aligned} \text{Speed} &= \frac{\text{distance}}{\text{time}} \\ &= \frac{200 \text{ meters}}{25 \text{ seconds}} \\ &= \frac{8 \text{ meters}}{1 \text{ second}} \\ &= 8 \text{ meters per second} \end{aligned}$$

Just as with other rates, we may instead write this as **8 meters/second**, **8 m/sec**, $8 \frac{\text{meters}}{\text{second}}$, or $8 \frac{\text{m}}{\text{sec}}$.

Speed problems in this chapter always refer to *average speed*, the ratio of *total* distance to *total* time. For example, in the problem above, Fiona probably wasn't always running the same speed, but her average speed was 8 meters per second.

PRACTICE | Answer each question about speed below.

122. The Beast Island activity blimp travels 90 miles in 3 hours. What is the blimp's average speed for the 3-hour trip in miles per hour?

122. _____

123. In one full day, the kudzu vine can grow 15 inches in length. How many inches per hour is this?

123. _____

124. Allison threw a baseball that traveled 145 feet in $2\frac{1}{2}$ seconds. What was the average speed of the ball in feet per second?

124. _____

125. Robbie ran 15 miles in $1\frac{2}{3}$ hours. What was Robbie's average speed in miles per hour?

125. _____

PRACTICE | Answer each question about speed below.

126. Erica runs $4\frac{1}{2}$ meters per second. At this speed, how many meters can she run in 45 seconds?

126. _____

127. Roland's remote-control boat has a maximum speed of 25 miles per hour. How many miles can Roland's boat travel in $\frac{1}{3}$ of an hour?

127. _____

128. An 8-kilometer hike up Yeti Mountain takes Grogg 3 hours. The hike back down on the same trail takes him just 2 hours. What is Grogg's average speed for his entire trip in kilometers per hour?

128. _____

129. Allison's rowing team rowed 5 miles in $\frac{3}{4}$ of an hour, then rowed another 3 miles in $\frac{7}{12}$ of an hour. What was their average speed for the entire trip in miles per hour?

129. _____

130. Annalise can walk 10 miles in 4 hours. At this rate, how many minutes does it take her to walk one mile?

130. _____

131. Urlich can paddle a kayak 5 miles per hour. At this speed, how
★ many minutes will it take him to travel $4\frac{1}{2}$ miles across Beast Bay?

131. _____

EXAMPLE | If 3 moles can dig 10 holes in 20 minutes, how many minutes will it take for 2 moles to dig 5 holes?

We assume that all of the moles work at the same rate.
We begin by considering how long it will takes 1 mole to dig 1 hole.

If 3 moles can dig 10 holes in 20 minutes, it will take 1 mole three times as long to complete the job. So, 1 mole can dig 10 holes in $3 \cdot 20 = 60$ minutes.

If 1 mole can dig 10 holes in 60 minutes, it will only take $\frac{1}{10}$ as long to dig 1 hole. So, 1 mole can dig 1 hole in $60 \cdot \frac{1}{10} = 6$ minutes.

Then, we use the 1-mole-1-hole time to find how long it takes 2 moles to dig 5 holes.

If 1 mole can dig 1 hole in 6 minutes, it will take 5 times as long to dig 5 holes. So, 1 mole can dig 5 holes in $6 \cdot 5 = 30$ minutes.

If 1 mole can dig 5 holes in 30 minutes, 2 moles working together will take half as long. So, 2 moles can dig 5 holes in $30 \cdot \frac{1}{2} = \mathbf{15\ minutes}$.

Rate problems can be pretty tricky when we have more than one worker completing a task.

For these problems, you may assume that each worker is working at the same rate!

PRACTICE | Fill in the blanks to answer each question below.

132. Five mowers can mow 4 football fields in 60 minutes. Use the reasoning below to figure out how many minutes it will take 3 mowers to mow 5 fields.

a. One mower can mow 4 football fields in _____ minutes.

b. One mower can mow 1 football field in _____ minutes.

c. Three mowers can mow 1 football field in _____ minutes.

d. Three mowers can mow 5 football fields in _____ minutes.

133. Six hoses can fill 5 buckets in 15 minutes. Use the reasoning below to figure out how many buckets can be filled by 3 hoses in 24 minutes.

a. Six hoses can fill 1 bucket in _____ minutes.

b. Six hoses can fill _____ buckets in 24 minutes.

c. Three hoses can fill _____ buckets in 24 minutes.

PRACTICE | Answer each question below.

134. Three gazellephants eat 3 buckets of peanuts in 3 minutes.

a. How many minutes would it take 6 gazellephants to eat 6 buckets of peanuts?

a. _____

b. How many buckets of peanuts can 6 gazellephants eat in 6 minutes?

b. _____

135. Twelve paint-bots can paint 20 houses in 9 hours.

a. How many paint-bots are needed to paint 5 houses in 9 hours?

a. _____

b. ★ How many houses can 9 paint-bots paint in 6 hours?

b. _____

136. Ten carpenters take 18 days to build 8 sheds.

a. How many days will it take 10 carpenters to build 12 sheds?

a. _____

b. ★ How many sheds can 3 carpenters build in 15 days?

b. _____

137. Six lumberjacks can chop 5 logs into firewood in 40 minutes.

a. ★ How many logs can 24 lumberjacks chop in 30 minutes?

a. _____

b. ★★ How many lumberjacks are needed to chop 6 logs in 72 minutes?

b. _____

EXAMPLE | How many fluid ounces are in 1 gallon?

We first convert gallons to quarts, then quarts to pints, and finally pints to ounces.

One gallon is 4 quarts.
One quart is 2 pints, so 4 quarts is $4 \cdot 2 = 8$ pints.
One pint is 16 fluid ounces, so 8 pints is $8 \cdot 16 = 128$ fl oz.

So, 1 gallon is **128 fl oz**.

PRACTICE | Answer each question below.

138. A mile is 5,280 feet. How many feet are there in 3 miles?

138. _____

139. There are 3 feet in one yard. One foot is 12 inches. How many inches are in 20 yards?

139. _____

140. There are 100 centimeters in a meter. One kilometer is 1,000 meters. How many centimeters are in 30 kilometers?

140. _____

141. How many minutes are in 1 week?

141. _____

142. ★ If you can jog 2 meters per second, how many kilometers can you jog in one hour?

142. _____

We can use **conversion factors** to convert between units.

A conversion factor is a fraction in which the numerator and denominator use different units to represent equal amounts.

For example, $\frac{1\ lb}{16\ oz}$ is a conversion factor because 1 pound equals 16 ounces. Similarly, since 16 ounces equals 1 pound, $\frac{16\ oz}{1\ lb}$ is a conversion factor.

Since the numerator and denominator of a conversion factor are equal, every conversion factor is equal to 1. So, we can multiply a measurement by a conversion factor to change its units without changing its value.

EXAMPLE | How many ounces is 6 pounds? How many pounds is 6 ounces?

To convert 6 pounds into ounces, we use the conversion factor $\frac{16\ oz}{1\ lb}$ to cancel the pounds units, as shown:

$$6\ lb = 6\ \cancel{lb} \cdot \frac{16\ oz}{1\ \cancel{lb}} = \textbf{96 oz}.$$

To convert 6 ounces into pounds, we use the conversion factor $\frac{1\ lb}{16\ oz}$ to cancel the ounces units, as shown:

$$6\ oz = 6\ \cancel{oz} \cdot \frac{1\ lb}{16\ \cancel{oz}} = \frac{6}{16}\ lb = \frac{3}{8}\ \textbf{lb}.$$

Check that your answers are reasonable. Since 1 pound is 16 ounces, 6 pounds is *more than* 16 ounces. Similarly, 6 ounces is *less than* 1 pound.

PRACTICE | Complete each calculation below to convert between units using the conversion factors given.

143. Use the equation below to find the number of fluid ounces in 15 cups, crossing out the units that cancel.

$$15\ cups = 15\ cups \cdot \frac{8\ fluid\ ounces}{1\ cup} = \underline{\hspace{2cm}}\ fluid\ ounces$$

144. Use the equation below to convert 60 ounces to pounds, crossing out the units that cancel.

$$60\ ounces = 60\ ounces \cdot \frac{1\ pound}{16\ ounces} = \underline{\hspace{2cm}}\ pounds$$

145. Use the equation below to convert from 150 feet per minute to feet per second, crossing out the units that cancel.

$$150\ feet\ per\ minute = \frac{150\ feet}{1\ minute} \cdot \frac{1\ minute}{60\ seconds} = \underline{\hspace{2cm}}\ feet\ per\ second$$

We want to choose the conversion factor that cancels the units that we are converting **from**.

For example, to convert **from** 20 quarts **to** gallons, we use the conversion factor $\frac{1 \text{ gallon}}{4 \text{ quarts}}$ so that the quarts units cancel:

$$20 \text{ quarts} \cdot \frac{1 \text{ gallon}}{4 \text{ quarts}} = 5 \text{ gallons}.$$

Below, we convert between some unusual units!

PRACTICE | Circle the conversion factor on the right you could use to complete each conversion below. Then, fill in each equation with the correct number.

146. 6 nanoseconds = _____ shakes

$\dfrac{1 \text{ shake}}{10 \text{ nanoseconds}}$ $\dfrac{10 \text{ nanoseconds}}{1 \text{ shake}}$

147. 30 fathoms = _____ feet

$\dfrac{1 \text{ fathom}}{6 \text{ feet}}$ $\dfrac{6 \text{ feet}}{1 \text{ fathom}}$

148. 2 nautical miles = _____ meters

$\dfrac{1 \text{ nautical mile}}{1,852 \text{ meters}}$ $\dfrac{1,852 \text{ meters}}{1 \text{ nautical mile}}$

149. 90 inches = _____ cubits

$\dfrac{1 \text{ cubit}}{18 \text{ inches}}$ $\dfrac{18 \text{ inches}}{1 \text{ cubit}}$

150. 10 joules = _____ kilocalories

$\dfrac{1 \text{ kilocalorie}}{4,184 \text{ joules}}$ $\dfrac{4,184 \text{ joules}}{1 \text{ kilocalorie}}$

151. 6 drams per fortnight = _____ ounces per fortnight

$\dfrac{1 \text{ ounce}}{16 \text{ drams}}$ $\dfrac{16 \text{ drams}}{1 \text{ ounce}}$

152. 200 bytes per jiffy = _____ bytes per second

$\dfrac{100 \text{ jiffies}}{1 \text{ second}}$ $\dfrac{1 \text{ second}}{100 \text{ jiffies}}$

PRACTICE | Draw a line to connect each rate conversion on the left to the conversion factor on the right you could multiply by to make the conversion.

153. 4 dollars per meter into cents per meter.

$$\frac{1{,}000 \text{ g}}{1 \text{ kg}}$$

154. 5 calories per gram into calories per kilogram.

$$\frac{5{,}280 \text{ ft}}{1 \text{ mi}}$$

155. 66 feet per minute into miles per minute.

$$\frac{1 \text{ mi}}{5{,}280 \text{ ft}}$$

156. 10 cents per meter into dollars per meter.

$$\frac{100 \text{ cents}}{1 \text{ dollar}}$$

157. 50,000 grains per kilogram into grains per gram.

$$\frac{1 \text{ dollar}}{100 \text{ cents}}$$

158. $\frac{1}{2}$ of a gallon per foot into gallons per mile.

$$\frac{1 \text{ kg}}{1{,}000 \text{ g}}$$

PRACTICE | Complete each rate conversion below using the conversion factors from the problems above.

159. 50,000 grains of rice weigh one kilogram. How many grains per gram is this?

159. _____ grains/gram

160. It takes $\frac{1}{2}$ of a gallon of sealant to cover one foot of driveway. How many gallons per mile of driveway is this?

160. _____ gal/mi

161. A bungee cord costs $4 per meter. How many cents per meter is this?

161. _____ ¢/m

162. Kelly Crab can scamper 66 feet per minute. What is Kelly's speed in miles per minute?

162. _____ mi/min

Conversions

EXAMPLE | A vine grows at a rate of 168 inches per week. How many feet per day is this?

Converting a rate can require converting multiple units in the numerator *and* denominator.

To begin, we write 168 inches per week as a fraction: $\frac{168 \text{ in}}{1 \text{ week}}$.

Then, we must convert from inches to feet, and from weeks to days.

To convert from inches to feet, we multiply $\frac{168 \text{ in}}{1 \text{ week}}$ by $\frac{1 \text{ ft}}{12 \text{ in}}$. The inches units cancel:

$$168 \text{ inches per week} = \frac{168 \text{ in}}{1 \text{ week}} \cdot \frac{1 \text{ ft}}{12 \text{ in}} = \frac{14 \text{ ft}}{1 \text{ week}} = 14 \text{ feet per week.}$$

Then, to convert from weeks to days, we multiply $\frac{14 \text{ ft}}{1 \text{ week}}$ by $\frac{1 \text{ week}}{7 \text{ days}}$. The week units cancel:

$$14 \text{ feet per week} = \frac{14 \text{ ft}}{1 \text{ week}} \cdot \frac{1 \text{ week}}{7 \text{ days}} = \frac{14 \text{ ft}}{7 \text{ days}} = \textbf{2 feet per day.}$$

— *or* —

We can perform both conversions at once:

$$168 \text{ inches per week} = \frac{168 \text{ in}}{1 \text{ week}} \cdot \frac{1 \text{ ft}}{12 \text{ in}} \cdot \frac{1 \text{ week}}{7 \text{ days}} = \frac{168 \text{ ft}}{84 \text{ days}} = \textbf{2 feet per day.}$$

You probably could have made this conversion without using conversion factors. However, using conversion factors allows us to organize our work for complicated conversions.

PRACTICE | Answer each question below.

163. A garden hose can fill a 30-gallon fish tank in 8 minutes. Use the equation below to determine how many fluid ounces flow from the hose every second, crossing out the units that cancel.

$$30 \text{ gallons in 8 minutes} = \frac{30 \text{ gal}}{8 \text{ min}} \cdot \frac{128 \text{ fl oz}}{1 \text{ gal}} \cdot \frac{1 \text{ min}}{60 \text{ sec}} = \underline{\hspace{2cm}} \text{ fluid ounces per second}$$

164. Joe uses a total of 3 cups of flour to make 2 batches of cookies. A cup of flour weighs 120 grams, and Joe makes 12 cookies in each batch. Use the equation below to determine how many grams of flour are in each cookie, crossing out the units that cancel.

$$3 \text{ cups for 2 batches} = \frac{3 \text{ cups}}{2 \text{ batches}} \cdot \frac{120 \text{ grams}}{1 \text{ cup}} \cdot \frac{1 \text{ batch}}{12 \text{ cookies}} = \underline{\hspace{2cm}} \text{ grams per cookie}$$

PRACTICE | Answer each question below.

165. Circle the two conversion factors below that you could multiply by to convert 200 feet per hour into inches per minute.

$$\frac{1 \text{ ft}}{12 \text{ in}} \qquad \frac{12 \text{ in}}{1 \text{ ft}} \qquad \frac{1 \text{ hr}}{60 \text{ min}} \qquad \frac{60 \text{ min}}{1 \text{ hr}}$$

166. Robbie the roverbot can travel 200 feet per hour. How many inches per minute is this?

166. _____

167. Circle the two conversion factors below that you could multiply by to convert 450 gallons per day into fluid ounces per hour.

$$\frac{1 \text{ gal}}{128 \text{ fl oz}} \qquad \frac{128 \text{ fl oz}}{1 \text{ gal}} \qquad \frac{1 \text{ day}}{24 \text{ hr}} \qquad \frac{24 \text{ hr}}{1 \text{ day}}$$

168. With machines running all day and night, Barry's Juice Factory presses 450 gallons of juice per day. How many fluid ounces of juice are pressed each hour?

168. _____

169. Circle the two conversion factors below that you could multiply by to convert 7 pounds per cup into ounces per teaspoon (tsp).

$$\frac{1 \text{ c}}{48 \text{ tsp}} \qquad \frac{48 \text{ tsp}}{1 \text{ c}} \qquad \frac{1 \text{ lb}}{16 \text{ oz}} \qquad \frac{16 \text{ oz}}{1 \text{ lb}}$$

170. Liquid mercury weighs about 7 pounds per cup. Using this estimate, how many ounces does a teaspoon of liquid mercury weigh?

170. _____

PRACTICE | Answer each question below.

171. There are 5,280 feet in one mile. Sam Snake slithers at a speed of 3 miles per hour. What is Sam's speed in feet per minute?

171. _____

172. A length of chain costs 15 cents per inch. What is the cost of this chain in dollars per yard?

172. _____

173. Emmanuel can jog 3 meters per second. What is his speed in kilometers per hour?

173. _____

174. There are 128 fluid ounces in a gallon. A pipe leaks 45 gallons of water every day. How many fluid ounces of water leak from the pipe per minute?

174. _____

175. ★ One serving of Beastie Bites cereal has 14 grams of sugar. A box of Beastie Bites contains 8 servings. There are about 28 grams in one ounce. Using this estimate, how many ounces of sugar are in a box of Beastie Bites?

175. _____

176. ★ We can use the conversion factor $\frac{254 \text{ cm}}{100 \text{ in}}$ to convert from inches to centimeters. A glacier moves 762 cm per day. At this speed, how many feet per week does the glacier move?

176. _____

nutes,
ear to

177. _____

178. Jeremy and Sarah each have a bag of lemons and limes.
★ The ratio of lemons to limes in Sarah's bag is 2:3. The ratio of lemons to limes in Jeremy's bag is 1:2. Sarah has 4 fruits for every 5 fruits that Jeremy has. After all 135 fruits are put in the same basket, what is the ratio of lemons to limes in the basket?

178. _____

179. A right triangle with a perimeter of 60 inches has sides whose lengths are in a 3:4:5 ratio. What is the area of the triangle, in square inches?

179. _____

180. Two giant tarantulemurs stand 70 meters apart on a huge web. They begin crawling directly toward each other. One crawls 2 meters per second, and the other crawls 3 meters per second. After how many seconds will the giant tarantulemurs meet?

180. _____

PRACTICE | Answer each question below.

181. The ratio of Jorble's height to Yorble's height is 5:9. If
★   Jorble is 10 inches shorter than Yorble, how many inches
    tall is Yorble?

181. _____

182. The ratio of the height of a rectangle to its perimeter is 2:19. What is
★   the ratio of the height of the rectangle to the width of the rectangle?

182. _____

183. The ratio of Alyssa's biking speed to her running speed is 5:2. What
★   is the ratio of the time it takes for Alyssa to bike to school to the time
    it takes her to run to school?

183. _____

PRACTICE | Answer each question below.

184. The ratio of blue cars to red cars in a parking lot is 2:3. After 15 blue
★   cars leave the lot, the ratio of blue cars to red cars is 1:2. How many
     red cars are in the parking lot?

184. _____

185. Everyone seated in the Orb Academy auditorium is either a student
★   or a teacher. The ratio of students to teachers is 5:1, and the ratio
     of filled seats to empty seats is 4:1. What is the ratio of teachers to
     empty seats in the auditorium?

185. _____

186. Two skateboards are 100 feet apart and rolling directly toward one
★   another, each at speed of 5 feet per second. Brody Bee starts at
     the front of one board and flies toward the other board at a speed
     of 7 feet per second. Whenever he reaches the front of a board,
     he reverses direction, flying back toward the other board at 7 feet
     per second. How far does Brody travel before the two skateboards
     collide, with Brody narrowly escaping the collision?

186. _____

CHAPTER 9
Decimals

Use this Practice book with
Guide 5C from BeastAcademy.com.

Recommended Sequence:

Book	Pages:
Guide:	70-77
Practice:	75-81
Guide:	78-82
Practice:	82-91
Guide:	83-101
Practice:	92-107

You may also read the entire chapter
in the Guide before beginning the
Practice chapter.

Decimals are another way to write fractions. Decimal place values are based on powers of 10, just like integers!

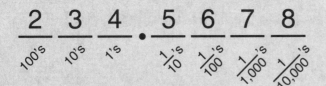

EXAMPLE | Write 0.125 as a fraction.

Review decimals basics in Chapter 11 of Practice 4D.

0.125 has a 1 in the tenths place, a 2 in the hundredths place, and a 5 in the thousandths place. So,

$$0.125 = \frac{1}{10} + \frac{2}{100} + \frac{5}{1,000}$$
$$= \frac{100}{1,000} + \frac{20}{1,000} + \frac{5}{1,000}$$
$$= \frac{125}{1,000}.$$

Then, we simplify: $\frac{125}{1,000} = \frac{1}{8}$.

— or —

We can write any decimal with three digits to the right of the decimal point as a number of thousandths. So, $0.125 = \frac{125}{1,000} = \frac{1}{8}$.

In this chapter, we usually write a zero to the left of the decimal point for numbers less than 1. However, decimals are sometimes written without the zero. For example, .2=0.2.

PRACTICE | Write the following decimals as fractions in simplest form. Use mixed numbers for fractions greater than 1.

1. 0.9 = _____

2. 0.327 = _____

3. 0.46 = _____

4. 6.128 = _____

5. 27.014 = _____

6. 505.0505 = _____

PRACTICE | Write the following fractions and mixed numbers as decimals.

7. $\frac{3}{10}$ = _____

8. $\frac{53}{100}$ = _____

9. $\frac{809}{1,000}$ = _____

10. $\frac{657}{100}$ = _____

11. $\frac{2,021}{1,000}$ = _____

12. $23\frac{71}{10,000}$ = _____

DECIMALS

EXAMPLE | Round 321.456 to the nearest tenth.

321.456 is between 321.<u>4</u> and 321.<u>5</u>.

321.4 ↑ 321.5
321.456

Since 321.456 is closer to 321.5 than to 321.4, we round 321.456 up to **321.5**.

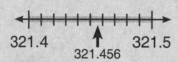
If necessary, include zeros at the end of a decimal when rounding.

For example, 2.397 rounds to 2.40 when rounded to the nearest hundredth.

PRACTICE | Round each number below to the indicated place value.

13. 1.43 rounds to _____
(nearest tenth)

14. 0.368 rounds to _____
(nearest tenth)

15. 77.7777 rounds to _____
(nearest hundredth)

16. 1.245 rounds to _____
(nearest hundredth)

17. 0.00317 rounds to _____
(nearest thousandth)

18. 8.2497 rounds to _____
(nearest thousandth)

PRACTICE | Solve each rounding problem below.

19. Insert a decimal point **between two digits** of each number below so that each resulting number has hundredths digit 6 when rounded to the nearest hundredth.

19659   56565   4565464

20. ★ When Grogg rounds the decimal below to the nearest hundredth, then rounds the result to the nearest tenth, he gets 0.4. Fill in the blanks to make the **smallest** possible value of Grogg's decimal.

0.3 __ __

EXAMPLE | Compute 29.34+7.8.

To add or subtract decimals, we stack the numbers so that the place values are lined up, filling any empty place values after a decimal point with 0's. Then, we compute the sum or difference the same way we do with whole numbers.

$$
\begin{array}{r}
29.34 \\
+ \ 7.80 \\
\hline
4
\end{array}
\qquad
\begin{array}{r}
\overset{1}{} \\
29.34 \\
+ \ 7.80 \\
\hline
14
\end{array}
\qquad
\begin{array}{r}
\overset{1\,1}{} \\
29.34 \\
+ \ 7.80 \\
\hline
7.14
\end{array}
\qquad
\begin{array}{r}
\overset{1\,1}{} \\
29.34 \\
+ \ 7.80 \\
\hline
37.14
\end{array}
$$

So, 29.34+7.8 = **37.14**.

PRACTICE | Compute each sum or difference below.

21. 2.35+6.91 = _____

22. 12.4+9.73 = _____

23. 5.9−2.65 = _____

24. 10.1−0.008 = _____

PRACTICE | Answer each question below.

25. What is the perimeter in centimeters of a triangle with side lengths 4.3 cm, 6.45 cm, and 9.9 cm?

25. _____

26. Alice ran 6.35 miles on a trail, then turned around and ran back to her starting point. Ben ran 4.4 miles of the trail, then ran back to his starting point. How many more miles did Alice run than Ben?

26. _____

27. What is the greatest possible difference that can be made by subtracting one of the numbers below from another?

0.23   3.54   1.0065   0.32   0.078   13.1   12.85

27. _____

EXAMPLE | Compute 0.32×10.

0.32 is 3 tenths and 2 hundredths, or $\frac{3}{10} + \frac{2}{100}$.

To multiply this sum by 10, we distribute the 10 as shown to the right.

So, $0.32 \times 10 = $ **3.2**.

$$0.32 \times 10 = \left(\frac{3}{10} + \frac{2}{100}\right) \times 10$$
$$= \left(\frac{3}{10} \times 10\right) + \left(\frac{2}{100} \times 10\right)$$
$$= 3 + \frac{2}{10}$$
$$= 3\frac{2}{10}$$
$$= 3.2.$$

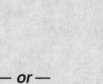

We use × for multiplication in this chapter to avoid confusing the multiplication dot with a decimal point.

— *or* —

For any two adjacent place values, the place value on the left is ten times the place value to its right. So, multiplying a number by 10 moves each digit to the next-larger place value to its left.

This is the same as moving the **decimal point** one place to the **right**. So, $0.32 \times 10 = $ **3.2**.

Multiplying any number by 10 shifts its decimal point one place to the right.

PRACTICE | Compute each product below.

28. $0.8 \times 10 = $ _____

29. $0.064 \times 10 = $ _____

30. $10 \times 3.7 = $ _____

31. $0.901 \times 10 \times 10 = $ _____

32. $10 \times 21.032 \times 10 = $ _____

33. $10 \times 3.001 \times 10 = $ _____

34. $12.3 \times 100 = $ _____

35. $1{,}000 \times 0.03405 = $ _____

36. $0.00002 \times 1{,}000 = $ _____

37. $0.040608 \times 10^5 = $ _____

PRACTICE | In each problem below, fill in the blank to make a true statement.

38. $1.821 \times$ _____ $= 18.21$

39. $0.007 \times$ _____ $= 7$

40. _____ $\times 0.065 = 6.5$

41. _____ $\times 10 = 0.3$

42. $100 \times$ _____ $= 71.3$

43. $10,000 \times$ _____ $= 345$

PRACTICE | Answer each question below.

44. If $0.02689 \times 10^n = 268.9$, then what is n?

44. $n =$ _____

45. Compute $0.043 \div 10$.

45. _____

46. ★ What is the smallest positive integer that can be multiplied by 0.002 to get an integer result?

46. _____

47. ★ What is the smallest positive integer that can be multiplied by 0.0025 to get an integer result?

47. _____

EXAMPLE | Compute 0.32×0.1.

0.32 is $\frac{3}{10} + \frac{2}{100}$, and 0.1 is $\frac{1}{10}$.

We use these fractions to compute 0.32×0.1, as shown to the right.

So, $0.32 \times 0.1 = \textbf{0.032}$.

$$0.32 \times 0.1 = \left(\frac{3}{10} + \frac{2}{100}\right) \times \frac{1}{10}$$
$$= \left(\frac{3}{10} \times \frac{1}{10}\right) + \left(\frac{2}{100} \times \frac{1}{10}\right)$$
$$= \frac{3}{100} + \frac{2}{1,000}$$
$$= 0.032.$$

— *or* —

For any two adjacent place values, the place value on the right is one tenth the place value to its left. So, multiplying a number by 0.1 moves each digit to the next-smaller place value to its right.

1's	$\frac{1}{10}$'s	$\frac{1}{100}$'s	$\frac{1}{1,000}$'s

$$\underline{\quad 0 \quad}.\underline{\quad 3 \quad}\underline{\quad 2 \quad} \quad \times 0.1$$
$$= \underline{\quad 0 \quad}.\underline{\quad 0 \quad}\underline{\quad 3 \quad}\underline{\quad 2 \quad}$$

This is the same as moving the **decimal point** one place to the **left**. So, $0.32 \times 0.1 = \textbf{0.032}$.

$$\underline{\quad 0 \quad}.\underline{\quad 3 \quad}\underline{\quad 2 \quad} \quad \times 0.1$$
$$= \underline{\quad 0 \quad}.\underline{\quad 0 \quad}\underline{\quad 3 \quad}\underline{\quad 2 \quad}$$

Multiplying any number by 0.1 shifts its decimal point one place to the left. This makes sense! Multiplying by a number less than 1 gives a product that is closer to zero than the number that we started with.

PRACTICE | Evaluate each expression below.

48. $1.5 \times 0.1 = $ _____

49. $0.1 \times 0.09 = $ _____

50. $0.1 \times 8 = $ _____

51. $340 \times 0.1 = $ _____

52. $12.34 \times 0.1 = $ _____

53. $6.5 \times 0.1 \times 0.1 = $ _____

54. $0.1 \times 0.046 \times 0.1 = $ _____

55. $831 \times 0.1 \times 0.1 \times 0.1 = $ _____

56. $0.0345 \div 10 = $ _____

57. ★ $0.0067 \div 0.1 = $ _____

PRACTICE | Compute each of the following powers of 0.1.

58. $(0.1)^2 = $ _____   **59.** $(0.1)^3 = $ _____   **60.** $(0.1)^6 = $ _____

PRACTICE | Evaluate each expression below.

61. $2.3 \times (0.1)^2 = $ _____

62. $0.3 \times 0.01 = $ _____

63. $(0.1)^3 \times 404 = $ _____

64. $6.7 \times 0.001 = $ _____

65. $(0.1)^2 \times 6.05 = $ _____

66. $8,500 \times 0.0001 = $ _____

67. $(0.1)^6 \times 4,000,000 = $ _____

68. $75.25 \times 0.00001 = $ _____

PRACTICE | In each problem below, fill in the blank with a decimal number to make a true statement.

69. $1.2 \times$ _____ $ = 0.012$

70. _____ $\times 312.5 = 0.3125$

71. $0.001 \times$ _____ $ = 0.03456$

72. _____ $\times 0.0001 = 0.25$

EXAMPLE | Compute 0.9×0.04.

We write each number as a fraction, multiply, then convert back to decimal form:

$$0.9 \times 0.04 = \frac{9}{10} \times \frac{4}{100}$$
$$= \frac{36}{1,000}$$
$$= \mathbf{0.036}.$$

— *or* —

Writing 0.9 as 9×0.1 and 0.04 as 4×0.01, we have

$$0.9 \times 0.04 = (9 \times 0.1) \times (4 \times 0.01)$$
$$= (9 \times 4) \times (0.1 \times 0.01)$$
$$= 36 \times 0.001$$
$$= \mathbf{0.036}.$$

PRACTICE | Evaluate each expression below.

73. $0.2 \times 0.3 = $ _____

74. $0.4 \times 0.8 = $ _____

75. $0.06 \times 0.11 = $ _____

76. $0.05 \times 0.3 = $ _____

77. $0.006 \times 0.12 = $ _____

78. $0.07 \times 0.8 \times 0.002 = $ _____

79. What positive number's square is 0.0049?

79. _____

80. ★ A rectangle has side lengths 0.A and 0.B inches, where A and B are each digits. If the perimeter of the rectangle is 1 inch, what is the largest possible area of the rectangle in square inches?

80. _____

EXAMPLE | Compute 0.9×0.04.

Instead of using one of the methods on the previous page, we consider how the decimal point moves when we multiply these numbers.

Since $0.9 = 9 \times 0.1$, multiplying by 0.9 is the same as multiplying by 9, then shifting the decimal point in the product 1 place to the left.

Similarly, since $0.04 = 4 \times 0.01$, multiplying by 0.04 is the same as multiplying by 4, then shifting the decimal point in the product 2 places to the left.

So, to compute 0.9×0.04 we can multiply 9×4, then shift the decimal point $1 + 2 = 3$ places to the left.

In other words, when multiplying any two numbers, we can just count the total number of digits to the right of the decimal point in both numbers to figure out where to place the decimal point in their product!

0.9 and 0.04 have a total of $1 + 2 = 3$ digits to the right of the decimal point.

$$0.9 \times 0.04$$
$$\underset{1}{} \quad \underset{2}{}$$

So, we move the decimal point in $9 \times 4 = 36$ so that there are 3 digits to the right of the decimal point.

$$0.036$$
$$\underset{3}{}$$

So, $0.9 \times 0.04 = \textbf{0.036}$.

PRACTICE | Place a decimal point in each product below to make each equation true.

81. $1.1 \times 1.1 = 121$

82. $10.1 \times 1.001 = 101101$

83. $33.34 \times 333.4 = 11115556$

84. $1{,}428.57 \times 0.007 = 999999$

PRACTICE | Evaluate each expression below.

85. $0.03 \times 0.8 = $ _____

86. $0.012 \times 9 = $ _____

87. $0.002 \times 0.0006 = $ _____

88. $0.0404 \times 0.07 = $ _____

Multiplying Decimals

Trailing zeros are the zeros at the end of a number that have no nonzero digits after them. Since trailing zeros after the decimal point do not change the value of a decimal, we usually don't write them.

For example, we usually write 0.2 instead of 0.20, 0.2000, 0.200000, and so on.

EXAMPLE | Compute 0.25×0.04.

We begin by multiplying $25 \times 4 = 100$. Then, we determine where to place the decimal point.

0.25 and 0.04 have a total of $2+2 = 4$ digits right of the decimal point.

$$0.\underset{2}{\underbrace{25}} \times 0.\underset{2}{\underbrace{04}}$$

So, we move the decimal point in 100 so that there are 4 digits to the right of the decimal point, **including** the trailing zeros.

$$0.\underset{4}{\underbrace{0100}}$$

After we have placed the decimal point, we can remove the trailing zeros.

$$0.01$$

So, $0.25 \times 0.04 = \mathbf{0.01}$.

> Be extra careful when the product of two numbers has trailing zeros!

PRACTICE | Compute each product below.

89. $0.2 \times 0.5 =$ _____

90. $0.06 \times 0.25 =$ _____

91. $0.075 \times 0.8 =$ _____

92. $0.00125 \times 0.032 =$ _____

93. ★ Not including trailing zeros, how many digits are to the right of the decimal point in the product below?

$$0.9 \times 0.8 \times 0.7 \times 0.6 \times 0.5 \times 0.4 \times 0.3 \times 0.2 \times 0.1$$

93. _____

94. ★ Not including trailing zeros, how many digits are to the right of the decimal point in the product $(0.3)^{15} \times (0.07)^{15}$?

94. _____

95. ★ Not including trailing zeros, how many digits are to the right of the decimal point in the product $(0.6)^{15} \times (0.05)^{15}$?

95. _____

PRACTICE | For the problems below, fill in each blank with a digit so that the equation is true and *no numbers have trailing zeros*.

96. $0.7 \times 0.\boxed{} = 0.5\boxed{}$

97. $10.\boxed{} \times 0.\boxed{} = 3.\boxed{}6$

98. $0.\boxed{} \times 0.2 = 0.\boxed{}$

99. $0.6 \times 2.\boxed{} = \boxed{}.\boxed{}$

100. $0.\boxed{} \times 4.\boxed{} = 0.\boxed{}3$

101. $0.0\boxed{} \times 0.0\boxed{} = 0.003$

102. $0.\boxed{}5 \times 0.0\boxed{} = 0.03$

103. $0.3\boxed{} \times 0.0\boxed{} = 0.007$

104. ★ $0.\boxed{}\boxed{}\boxed{} \times 0.\boxed{} = 0.7$

105. ★ $1.\boxed{} \times 0.\boxed{}\boxed{}\boxed{} = 0.04$

EXAMPLE | Compute $0.0004 \times 5,000$.

Since $0.0004 = 4 \times 0.0001$, multiplying by 0.0004 is the same as multiplying by 4, then shifting the decimal point in the product 4 places to the **left**.

Similarly, since $5,000 = 5 \times 1,000$, multiplying by 5,000 is the same as multiplying by 5, then shifting the decimal point in the product 3 places to the **right**.

So, to compute $0.0004 \times 5,000$, we can multiply $4 \times 5 = 20$, then shift the decimal point 4 places to the left and 3 places to the right. This is the same as shifting the decimal point 1 place to the left:

$$2.0$$

So, $0.0004 \times 5,000 = 2.0 = \mathbf{2}$.

> The number of trailing zeros in an integer tells us how far to move the decimal point to the right.
>
> Since 5,000 has 3 trailing zeros, multiplying by 5,000 moves the decimal point 3 places to the right.

PRACTICE | Evaluate each expression below.

106. $200 \times 0.009 = $ _____

107. $0.04 \times 13,000 = $ _____

108. $0.00025 \times 800 = $ _____

109. $300 \times 0.03 \times 0.003 = $ _____

110. $40 \times 0.012 \times 1,000 = $ _____

111. $0.025 \times 0.04 \times 2,000 = $ _____

112. ★ Not including trailing zeros, how many digits are to the right of the decimal point in the product $(300)^{10} \times (0.002)^{10}$?

112. _____

113. ★ Compute the product $(400)^{10} \times (0.0025)^{10}$.

113. _____

PRACTICE | Answer each question below.

114. A toy brick is 0.64 centimeters tall. What is the height in centimeters of a stack of 40 of these toy bricks?

114. _____

115. What is the volume in cubic centimeters of a cube whose edges are 0.4 centimeters long?

115. _____

116. One inch is equal to exactly 2.54 centimeters. What is the height in centimeters of a gentlebug who is 0.3 inches tall?

116. _____

117. Timmy earns $12.25 per hour as a lifeguard. How many dollars will Timmy earn after 6 hours of lifeguarding?

117. _____

118. ★ The product of 0.125 and a is 0.01. What is a?

118. $a =$ _____

119. ★ What is the total area in square meters of the two rectangles below?

119. _____

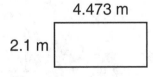

4.473 m

2.1 m

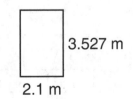

3.527 m

2.1 m

The goal in a **Pyramid Descent** puzzle is to find a path of touching blocks so that the product of the numbers in the path equals the number shown above the pyramid.

Each path moves from the top to the bottom of the pyramid, touching one block per row.

0.07

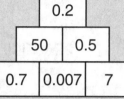

EXAMPLE | Complete the Pyramid Descent puzzle to the right.

Ignoring the decimal points and trailing zeros in each block, the top block contains a 2, each block in the middle row contains a 5, and each block in the bottom row contains a 7. So, ignoring decimal points and trailing zeros, every path has a product of 2×5×7 = 70.

The product we seek is 0.07. To go from 70 to 0.07, we move the decimal point three places to the left. The direction in which we move the decimal point and the number of times we move it is determined by the number of trailing zeros and the number of digits after the decimal point in each path.

0.07

For example, in the path shown to the right, there are a total of 2 digits after the decimal point, which move the decimal point 2 places to the left. There is also 1 trailing zero, which moves the decimal point in the product 1 place to the right. All together, the decimal point moves 1 place to the left, so this path has a product of 7.

$0.2 \times 50 \times 0.7 = 7.0 = 7$

We consider the placement of the decimal point for each remaining path. The only path whose product is 0.07 is shown with the circled numbers below.

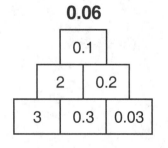

$0.2 \times 50 \times 0.007 = 0.070 = \mathbf{0.07}$

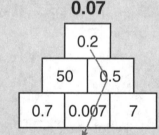

$0.2 \times 0.5 \times 0.007 = 0.00070 = 0.0007$

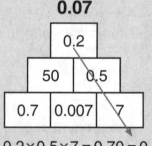

$0.2 \times 0.5 \times 7 = 0.70 = 0.7$

PRACTICE | Complete each Pyramid Descent puzzle below.

120.

0.06

0.1		
2	0.2	
3	0.3	0.03

121.

0.008

0.2		
0.2	0.02	
20	0.02	2

122.

0.3

0.2		
30	0.03	
0.5	0.05	5

PRACTICE | Complete each Pyramid Descent puzzle below.

123.

8.1

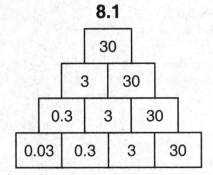

124.

0.1001

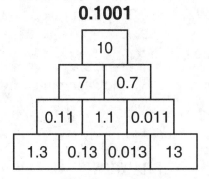

125.

0.01

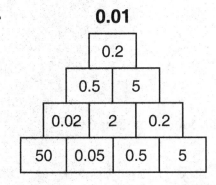

126.

0.024

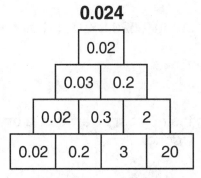

127.

0.012

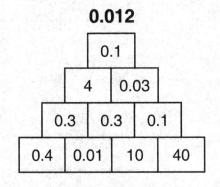

128.

0.6

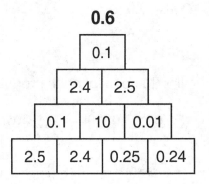

If the **exact** answer to a difficult decimal computation is needed, it's usually best to use a calculator.

However, for many computations, an estimate is all you need.

You will not need a calculator for any of the problems in this book.

EXAMPLE | Estimate the product of 0.032 and 11.93.

Since 0.032 is about 0.03, and 11.93 is about 12, we estimate that $0.032 \times 11.93 \approx 0.03 \times 12 = \textbf{0.36}$.

In fact, $0.032 \times 11.93 = 0.38176$. The difference between 0.38176 and 0.36 is 0.02176, which is small compared to 0.36. So, 0.36 is a reasonable estimate. ✓

You may have arrived at a different estimate. Any estimate between 0.3 and 0.4 is reasonable.

PRACTICE | Use estimation to answer each question below.

129. Is the product 0.0198×407.6 closer to 8 or to 80?

129. _____

130. Is the product 523.9×0.0031 closer to 1.5 or to 15?

130. _____

131. Circle the number below that is equal to 0.068×297.4 without computing the exact product.

     2.2232   20.2232   30.2232   180.2232   202.232

132. Circle the product below that is equal to 0.9594 without computing any of the products.

     2.46×0.039 0.82×11.7 0.018×5.33 0.234×4.1

133. Draw a line to connect each product below to its correct location on the number line.

    50.03×0.0819 0.68×0.392 0.01207×42.9 0.054×102.29

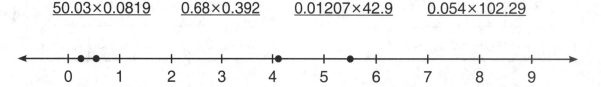

PRACTICE | Fill each circle below with < or > to indicate which expression is greater.

134. 0.9914×0.52 ◯ 0.52

135. 3.911×0.049 ◯ 0.2

136. 9.008×0.7134 ◯ 6.3

137. 0.852 ◯ 98.05×0.00852

138. 0.029×0.38 ◯ 2.05×0.0061

139. 3.208×0.768 ◯ 4.95×0.3809

PRACTICE | Use estimation to answer each question below.

140. The product 0.0625×48 is equal to what integer?

140. _____

141. How many multiples of 10 are between 4.104×5.724 and 99?

141. _____

142. Circle the two products below that are equal.

0.028×7.03 $\qquad$ 0.982×0.41 $\qquad$ 0.62×1.571 $\qquad$ 0.76×0.259

143. Ralph uses a calculator to compute 0.0058×21.3 and gets an answer of 1.2354. Is his answer reasonable? Explain.

EXAMPLE | Write the fraction $\frac{7}{20}$ as a decimal.

We can easily write a fraction as a decimal when the denominator of the fraction is a power of 10.

Multiplying the numerator and denominator of $\frac{7}{20}$ by 5 gives an equivalent fraction with denominator 100.

$$\frac{7}{20} \overset{\times 5}{\underset{\times 5}{=}} \frac{35}{100}$$

So, $\frac{7}{20} = \frac{35}{100} = \mathbf{0.35}$.

PRACTICE | Write each fraction below as a decimal.

144. $\frac{13}{50} = $ _____

145. $\frac{3}{4} = $ _____

146. $\frac{8}{5} = $ _____

147. $\frac{7}{25} = $ _____

148. $\frac{51}{20} = $ _____

149. $\frac{5}{8} = $ _____

150. $\frac{9}{125} = $ _____

151. $\frac{3}{40} = $ _____

152. $\frac{33}{8} = $ _____

153. ★ When the fraction $\frac{13}{2^{99} \times 5^{101}}$ is written as a decimal, what is the sum of all the digits to the right of the decimal point?

153. _____

EXAMPLE | Write the fraction $\frac{7}{20}$ as a decimal.   *Fractions to Decimals*

We could use the method on the previous page. Or, we can use long division to convert fractions into decimals.

Computing $\frac{7}{20} = 7 \div 20$ using long division is easier to think about if we write 7 as 7.0. Then, we can ignore the decimal point for a moment and think of $20\overline{)7.0}$ as $20\overline{)70}$.

$20\overline{)7.0}$

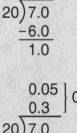

Review long division in the Division chapter of Beast Academy 4B.

Check out an alternative long division algorithm on page 105.

Since $20 \times 3 = 60$, we know that 20 goes into 70 3 times with 10 left over. Dividing 20 into 7.0 is similar. Since $20 \times 0.3 = 6.0$, we know that 20 goes into 7.0 0.3 times with 1.0 left over.

$$\begin{array}{r} 0.3 \\ 20\overline{)7.0} \\ -6.0 \\ \hline 1.0 \end{array}$$

Next, we divide 20 into 1.0. This is easier to think about if we write 1.0 as 1.00 and consider dividing 20 into 100.

Since $20 \times 5 = 100$, we know 20 goes into 100 5 times with 0 left over. Dividing 20 into 1.00 is similar. Since $20 \times 0.05 = 1.00$, we know 20 goes into 1.00 0.05 times with 0 left over.

$$\begin{array}{r} 0.05 \\ 0.3 \quad \Big| \, 0.35 \\ 20\overline{)7.0} \\ -6.0 \\ \hline 1.00 \\ -1.00 \\ \hline 0 \end{array}$$

So, $\frac{7}{20} = 7 \div 20 = 0.3 + 0.05 = \mathbf{0.35}$.

Check: $20 \times 0.35 = 7.0 = 7.$ ✓

PRACTICE | Use long division to write each fraction below as a decimal.

154. Fraction: $\frac{4}{5}$ $5\overline{)4.0}$

Decimal: _____

155. Fraction: $\frac{3}{4}$ $4\overline{)3.0}$

Decimal: _____

156. Fraction: $\frac{1}{40}$ $\overline{)}$

Decimal: _____

157. Fraction: $\frac{15}{8}$ $\overline{)}$

Decimal: _____

Print more practice problems at www.BeastAcademy.com

Fractions to Decimals

If a fraction can be written with a denominator that is a power of 10, then it can be written as a **terminating decimal**. This means that its decimal digits do not go on and on forever (not including trailing zeros).

For example, $\frac{63}{64}$ can be written as $\frac{984,375}{1,000,000} = 0.984375$. However, not all decimals terminate.

EXAMPLE | Write the fraction $\frac{9}{11}$ as a decimal.

We cannot write $\frac{9}{11}$ as an equivalent fraction with a denominator that is a power of 10.

However, we can convert $\frac{9}{11}$ to a decimal using long division.

11 goes into 9.0
0.8 times with 0.2 left over.

Then, 11 goes into 0.20
0.01 times with 0.09 left over.

Then, 11 goes into 0.090
0.008 times with 0.002 left over.

Then, 11 goes into 0.0020
0.0001 times with 0.0009 left over.

$$
\begin{array}{r}
0.8 \\
11\overline{)9.0} \\
-8.8 \\
\hline
0.2
\end{array}
\qquad
\begin{array}{r}
0.01 \\
0.8 \\
11\overline{)9.0} \\
-8.8 \\
\hline
0.20 \\
-0.11 \\
\hline
0.09
\end{array}
\qquad
\begin{array}{r}
0.008 \\
0.01 \\
0.8 \\
11\overline{)9.0} \\
-8.8 \\
\hline
0.20 \\
-0.11 \\
\hline
0.090 \\
-0.088 \\
\hline
0.002
\end{array}
\qquad
\begin{array}{r}
0.0001 \\
0.008 \\
0.01 \\
0.8 \\
11\overline{)9.0} \\
-8.8 \\
\hline
0.20 \\
-0.11 \\
\hline
0.090 \\
-0.088 \\
\hline
0.0020 \\
-0.0011 \\
\hline
0.0009
\end{array}
$$

We notice a pattern! The remainders alternate between decimals ending in 2 and 9. When we divide these remainders by 11, the quotients alternate between decimals ending in 8 and 1. So, $\frac{9}{11} = 0.818181\ldots$.

Decimals whose digits repeat in a pattern are called **repeating decimals**. We usually write repeating decimals by drawing a bar over the digits that repeat. For example, we write 0.444... as $0.\overline{4}$, we write 0.1555... as $0.1\overline{5}$, and we write 0.0454545... as $0.0\overline{45}$.

So, $\frac{9}{11} = 0.818181\ldots = \mathbf{0.\overline{81}}$.

PRACTICE | Use long division to write each fraction below as a repeating decimal.

158. Fraction: $\frac{1}{3}$

$$3\overline{)1.0}$$

Decimal: _____

159. Fraction: $\frac{5}{6}$

$$6\overline{)5.0}$$

Decimal: _____

PRACTICE | Write each fraction below as a decimal using long division.

160. $\frac{4}{9} = $ _____

161. $\frac{2}{15} = $ _____

162. $\frac{13}{12} = $ _____

PRACTICE | Answer each question below.

163. When $\frac{3}{11}$ is written as a decimal, what is the 100th digit to the right of the decimal point?

163. _____

164. When $\frac{1}{33}$ is written as a decimal, what is the 99th digit to the right of the decimal point?

164. _____

165. Only fractions that can be written with a denominator that is a power of 10 can be written as terminating decimals. Circle each fraction below that can be written as a terminating decimal.

$\frac{1}{25}$ $\frac{1}{30}$ $\frac{1}{91}$ $\frac{1}{125}$ $\frac{1}{55}$ $\frac{1}{32}$

166. If the fraction $\frac{a}{b}$ is in simplest form and can be written as a terminating decimal, then what must be true about the prime factorization of b?

Knowing the decimal form of one fraction can help us figure out the decimal forms of other fractions.

EXAMPLE | Use the fact that $\frac{1}{3} = 0.\overline{3}$ to write $\frac{1}{30}$ as a decimal.

Notice that $\frac{1}{30} = \frac{1}{10} \times \frac{1}{3}$.

Since $\frac{1}{10} = 0.1$ and $\frac{1}{3} = 0.\overline{3}$, we have $\frac{1}{30} = 0.1 \times 0.\overline{3}$.

Multiplying a number by 0.1 moves the decimal point one place to the left.

So, $\frac{1}{30} = \frac{1}{10} \times \frac{1}{3} = 0.1 \times 0.\overline{3} = \mathbf{0.0\overline{3}}$.

> Be careful not to confuse $0.0\overline{3} = 0.033333...$ with $0.\overline{03} = 0.030303...$

PRACTICE | Answer each question below.

167. Use the fact that $\frac{1}{8} = 0.125$ to write $\frac{1}{80}$ as a decimal.

167. $\frac{1}{80} = $ _____

168. Multiply $\frac{1}{3} = 0.\overline{3}$ by 2 to write $\frac{2}{3}$ as a decimal.

168. $\frac{2}{3} = $ _____

169. Divide $\frac{1}{3} = 0.\overline{3}$ by 3 to write $\frac{1}{9}$ as a decimal.

169. $\frac{1}{9} = $ _____

170. Use your answer to the previous problem to write each multiple of $\frac{1}{9}$ below as a decimal.

$\frac{2}{9} = $ _____ $\frac{3}{9} = $ _____ $\frac{4}{9} = $ _____ $\frac{5}{9} = $ _____ $\frac{6}{9} = $ _____ $\frac{7}{9} = $ _____ $\frac{8}{9} = $ _____

171. ★ Use the strategies above to help you fill in the blanks in the statements below.

$\frac{1}{9} = $ _____, so $\frac{1}{90} = $ _____. Therefore, $\frac{1}{45} = \frac{2}{90} = $ _____, and $\frac{1}{15} = \frac{6}{90} = $ _____.

EXAMPLE | Use the fact that $\frac{1}{3}=0.\overline{3}$ to write $\frac{1}{33}$ as a decimal.

We notice that $\frac{1}{33}=\frac{1}{3}\times\frac{1}{11}$. Multiplying by $\frac{1}{11}$ is the same as dividing by 11, so we divide $0.\overline{3}$ by 11.

Thinking of $0.\overline{3}$ as $0.\overline{33}$, with repeating blocks of 33, makes dividing by 11 easier. We know that $0.33\div11=0.03$, $0.3333\div11=0.0303$, and $0.333333\div11=0.030303$.

From this pattern, we see that $(0.333333...)\div11=0.030303...=0.\overline{03}$.

So, $\frac{1}{33}=\frac{1}{3}\times\frac{1}{11}=0.\overline{33}\times\frac{1}{11}=0.\overline{33}\div11=\mathbf{0.\overline{03}}$.

PRACTICE | Answer each question below.

172. Divide $\frac{1}{33}=0.\overline{03}$ by 3 to write $\frac{1}{99}$ as a decimal.

172. $\frac{1}{99}=$ _____

173. Use your answer to the previous problem to write each multiple of $\frac{1}{99}$ below as a decimal.

$\frac{2}{99}=$ _____ $\qquad$ $\frac{7}{99}=$ _____ $\qquad$ $\frac{17}{99}=$ _____ $\qquad$ $\frac{49}{99}=$ _____ $\qquad$ $\frac{9}{99}=\frac{1}{11}=$ _____

174. Write $0.\overline{36}$ as a fraction in simplest form.

174. $0.\overline{36}=$ _____

175. ★ Use the strategies above to help you fill in the blanks in the statements below.

$\frac{1}{3}=0.\overline{333}$, so $\frac{1}{333}=\frac{1}{3}\times\frac{1}{111}=$ _____, and $\frac{1}{999}=$ _____. Therefore, $\frac{1}{37}=\frac{27}{999}=$ _____.

PRACTICE | Answer each question below.

176. Write $\frac{100}{99}$ as a decimal.

176. $\frac{100}{99} =$ _____

177. Convert $\frac{7}{11}$ to 99ths to write $\frac{7}{11}$ as a decimal.

177. $\frac{7}{11} =$ _____

178. Convert $\frac{4}{33}$ to 99ths to write $\frac{4}{33}$ as a decimal.

178. $\frac{4}{33} =$ _____

179. Use the fact that $\frac{7}{9} = 0.\overline{7}$ to write $\frac{7}{90}$ as a decimal.

179. $\frac{7}{90} =$ _____

180. Circle the number below that is equal to $4 \times 0.0\overline{7}$.

$0.2\overline{8}$ $0.\overline{28}$ $0.2\overline{9}$ $0.\overline{29}$ $0.3\overline{1}$ $0.\overline{31}$

181. Convert $\frac{13}{18}$ to 90ths to write $\frac{13}{18}$ as a decimal.

181. $\frac{13}{18} =$ _____

EXAMPLE | Use the fact that $\frac{1}{6}=\frac{1}{2}-\frac{1}{3}$ to write $\frac{1}{6}$ as a decimal.

Since $\frac{1}{2}=0.5$ and $\frac{1}{3}=0.\overline{3}$, we try to subtract $0.5-0.\overline{3}$.

We line up the decimal points and write 0's after the 5 in 0.5.
To subtract, we need to take a tenth from the tenths place in 0.5
to make 10 hundredths. Then, we take a hundredth to make 10
thousandths, and so on. Then, we can subtract as shown below.

$$
\begin{array}{r}
0.5000... \\
-\ 0.3333... \\
\end{array}
\qquad
\begin{array}{r}
{}^{4\,10}\\
0.\cancel{5}000... \\
-\ 0.3333... \\
\end{array}
\qquad
\begin{array}{r}
{}^{4\,9\,9\,9\,...}\\
0.\cancel{5}\cancel{0}\cancel{0}0... \\
-\ 0.3333... \\
\end{array}
\qquad
\begin{array}{r}
{}^{4\,9\,9\,9\,...}\\
0.\cancel{5}\cancel{0}\cancel{0}\cancel{0}... \\
-\ 0.3333... \\
\hline
0.1666... \\
\end{array}
$$

So, $\frac{1}{6}=0.1\overline{6}$.

PRACTICE | Answer each question below.

182. Use the fact that $\frac{1}{2}+\frac{1}{3}=\frac{5}{6}$ to write $\frac{5}{6}$ as a decimal.

182. $\frac{5}{6}=$ _____

183. Use the fact that $\frac{1}{3}+\frac{1}{4}=\frac{7}{12}$ to write $\frac{7}{12}$ as a decimal.

183. $\frac{7}{12}=$ _____

184. Use the fact that $\frac{2}{3}+\frac{1}{4}=\frac{11}{12}$ to write $\frac{11}{12}$ as a decimal.

184. $\frac{11}{12}=$ _____

185. Use the fact that $\frac{2}{3}-\frac{1}{4}=\frac{5}{12}$ to write $\frac{5}{12}$ as a decimal.

185. $\frac{5}{12}=$ _____

186. Use the fact that $\frac{1}{3}-\frac{1}{4}=\frac{1}{12}$ to write $\frac{1}{12}$ as a decimal.

186. $\frac{1}{12}=$ _____

In a **Frac-Turn** puzzle, each fraction begins a path on a grid. The path is directed by arrows until it leaves the grid or creates a loop. The goal is to fill some or all of the squares in the grid with digits so that the path from each fraction traces its decimal form, with no leading zeros to the left of the decimal point. Below is a solved example, with each path traced.

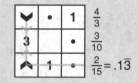

EXAMPLE | Solve the Frac-Turn puzzle to the right.

We begin by converting each fraction to a decimal.

$$\frac{1}{50} = .02 \qquad \frac{7}{33} = \frac{21}{99} = .\overline{21} \qquad \frac{18}{5} = 3\frac{3}{5} = 3.6 \qquad \frac{4}{33} = \frac{12}{99} = .\overline{12}$$

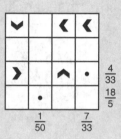

The path for $\frac{18}{5} = 3.6$ has only one empty square after the decimal point. So, we must place a 6 in the bottom-left square, and a 3 in one of the other empty squares in the bottom row. The 3 cannot go in the bottom-right square because this square is in the path of $\frac{7}{33} = .\overline{21}$. So, we fill in the 3 as shown to the right.

Next, the paths of $\frac{4}{33} = .\overline{12}$ and $\frac{7}{33} = .\overline{21}$ both contain a loop.

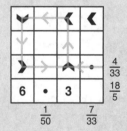 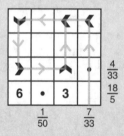

No square within either loop can contain a digit other than 1 or 2. So, we must place the 0 in $\frac{1}{50} = .02$ as shown to the right. Then, we place the 2 in the square just above the 0.

Finally, we place a 1 in the only square in the loop that correctly completes the paths of $\frac{4}{33} = .\overline{12}$ and $\frac{7}{33} = .\overline{21}$. Each fraction's path traces its decimal form, so we are done.

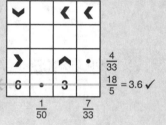

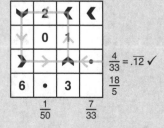 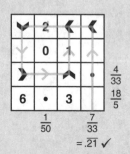 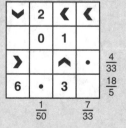

Remember, we don't write a leading zero before the decimal point in these puzzles!

PRACTICE | Solve each Frac-Turn puzzle below.

187. $\frac{3}{5}$

$\frac{1}{4}$

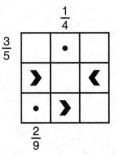

$\frac{2}{9}$

188. $\frac{3}{50}$

$\frac{1}{6}$ $\frac{8}{3}$

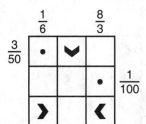

$\frac{1}{100}$

189.

$\frac{3}{5}$
$\frac{2}{3}$

$\frac{1}{6}$ $\frac{1}{10}$

190. $\frac{13}{4}$ $\frac{21}{4}$

$\frac{6}{5}$

191. $\frac{65}{9}$

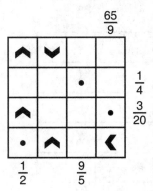

$\frac{1}{4}$
$\frac{3}{20}$

$\frac{1}{2}$ $\frac{9}{5}$

192. $\frac{23}{3}$

$\frac{9}{25}$

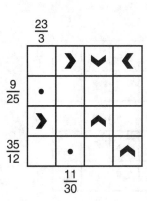

$\frac{35}{12}$

$\frac{11}{30}$

PRACTICE | Solve each Frac-Turn puzzle below.

193.

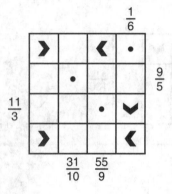

194.

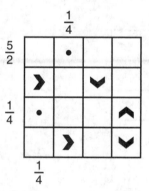

195. ★

196. ★

197.

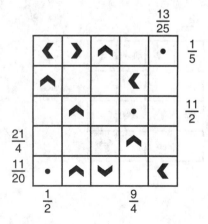

198.

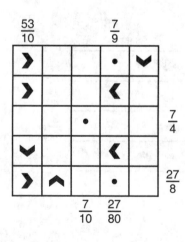

PRACTICE | Solve each Frac-Turn puzzle below.

199.

200.

201. Complete each of the following to solve the Frac-Turn puzzle below:

- Write each fraction on the left in its correct location on the outside of the grid.
- Place **all ten digits** (from 0-9) once each on the grid so that each fraction's path correctly traces its decimal form.

<u>Fractions</u>

$\frac{29}{40}$ $\frac{1}{6}$

$\frac{21}{4}$ $\frac{2}{11}$

$\frac{9}{4}$ $\frac{9}{11}$

$\frac{9}{10}$ $\frac{1}{15}$

$\frac{24}{5}$

Many decimal quotients can be converted into integer quotients that are easier to compute.

EXAMPLE | Compute $4.8 \div 0.16$.

We can use long division to divide 0.16 into 4.8.

Or, we can convert this into a quotient of integers. This is easiest to see when we write the quotient as a fraction:

$$4.8 \div 0.16 = \frac{4.8}{0.16}.$$

Multiplying the numerator and denominator of $\frac{4.8}{0.16}$ by 100 gives an equivalent fraction whose numerator and denominator are both integers.

$$\frac{4.8}{0.16} \xrightarrow{\times 100} \frac{480}{16} \xleftarrow{\times 100}$$

Then, we compute $480 \div 16 = \mathbf{30}$.

PRACTICE | Express each quotient below as a fraction or mixed number in simplest form.

202. $0.05 \div 1.25 =$ _____

203. $0.4 \div 2.2 =$ _____

204. $3.03 \div 0.6 =$ _____

205. $3.5 \div 70.707 =$ _____

PRACTICE | Express each quotient below as a whole number or decimal.

206. $\frac{0.35}{0.05} =$ _____

207. $\frac{2.4}{0.03} =$ _____

208. $\frac{0.032}{0.8} =$ _____

209. $4.9 \div 0.14 =$ _____

210. $0.002 \div 0.04 =$ _____

211. $0.6 \div 4.5 =$ _____

Extra! The Traditional Algorithm for Long Division

Many people use an algorithm (a set of steps) for long division that is slightly different from the one we have used in Beast Academy.

EXAMPLE | Write the fraction $\frac{15}{37}$ as a decimal.

Traditional Algorithm:

In the traditional algorithm, we find the digit in each place value of the quotient, working from the largest place value to the smallest until the decimal repeats or terminates.

```
        tens
        ones
        tenths
        hundredths...
          0.
  37) 15.0
```

Step 1:
We locate the first nonzero digit in our quotient. Since 37 is greater than 15, the quotient is less than 1. So, the ones digit of our quotient is 0. We write a 0 in the ones place of our quotient, aligning the decimal point and all of the place values of our quotient with the place values of the dividend, 15. Then, we write 15 as 15.0, which is 150 tenths.

```
         0.4
  37) 15.0
      -14.8
        0.2
```

Step 2:
To find the tenths digit of our quotient, we divide 150 *tenths* by 37. 37 goes into 150 a total of 4 times, so there are 4 tenths in the quotient. We write a 4 in the tenths place of our quotient. $4 \times 37 = 148$, so our remainder is $150 - 148 = 2$ tenths (0.2). If we keep our place values and decimal points aligned, we can think of all the work in terms of whole numbers.

```
         0.40
  37) 15.0
      -14.8
        0.20
```

Step 3:
Next, we find the hundredths digit of the quotient. We write a 0 in the hundredths place of 0.2 to get 20 hundredths (0.20). Since 37 goes into 20 0 times, we write a 0 in the hundredths place of our quotient.

```
         0.405
  37) 15.0
      -14.8
        0.200
      -0.185
        0.015
```

Step 4:
Next, we find the thousandths digit of the quotient. We write a 0 in the thousandths place of 0.20 to get 200 thousandths (0.200). Since 37 goes into 200 5 times, we write a 5 in the thousandths place of our quotient. $37 \times 5 = 185$, so our remainder is $200 - 185 = 15$ thousandths (0.015).

```
         0.405
  37) 15.0
      -14.8
        0.200
      -0.185
        0.0150
```

Step 5:
Next, we find the ten-thousandths digit of the quotient. We write a 0 in the ten-thousandths place of 0.015 to get 150 ten-thousandths (0.0150). But wait, we've done this before! In Step 2, we divided 150 by 37. Now, we are dividing 150 by 37 again. We get the same quotient (4) and remainder (2) as we did before, and our digits begin to repeat.

So, $\frac{15}{37} = 0.405405405...$, or **$0.\overline{405}$**.

There are many variations on the long division algorithm. We encourage you to find one that works best for you. However, for many problems that require long division, it's fine to use a calculator.

PRACTICE | Solve each problem below.

212. Not including trailing zeros, what is the rightmost digit when $\left(\frac{1}{2}\right)^{99}$ is written as a decimal?

212. _____

213. When $\left(\frac{1}{5}\right)^{10}$ is written as a decimal, how many zeros are to the right of the decimal point before the first nonzero digit?

213. _____

214. Potatoes cost $0.67 per pound. What is the smallest whole number of dollars you will need to buy 6.375 pounds of potatoes?

214. _____

215. What is the smallest positive number that 0.008 can be multiplied by ★ to get a decimal product that has exactly 3 digits after the decimal point, not including trailing zeros?

215. _____

PRACTICE | Solve each problem below.

216. Order $0.\overline{05}$, $0.\overline{050}$, 0.05, and $0.0\overline{5}$ from least to greatest on the lines below.

_____   _____   _____   _____

217. Express the sum $0.\overline{1} + 0.0\overline{1} + 0.00\overline{1}$ as a fraction in simplest form.

217. _____

218. ★ What is the smallest positive integer you can multiply by $\frac{1}{8!}$ so that the result can be expressed as a terminating decimal?

218. _____

219. ★ Express $\frac{0.0\overline{3}}{0.\overline{2}}$ as a decimal.

219. _____

220. ★ ✎ Grogg adds three copies of $0.\overline{3}$ and gets $0.\overline{9}$, as shown below. What simpler number is equal to $0.\overline{9}$? Explain.

$$\begin{array}{r} 0.3333... \\ 0.3333... \\ + \ 0.3333... \\ \hline 0.9999... \end{array}$$

HINTS
For Selected Problems

Below are hints to every problem marked with a ★.
Work on the problems for a while before looking at the hints.
The hint numbers match the problem numbers.

7. Where are the endpoints of each new diagonal?

14. This one is out of this world!

15. It might help to look at this sequence in the mirror.

16. What do the letters in the top row have in common?

17. Too good of a hint might give it away.
So a cryptic rhyme is all that we'll say.

23. It might help to write the whole numbers as fractions.

28. What is the repeating pattern in this sequence?

33. What happens when you raise a negative number to an even power? An odd power?

34. It might help to write the whole numbers as fractions.

48. How can you use the sum of the terms to find the middle term in an arithmetic sequence?

62. What expression gives the kth term for any number k?

73. What sequence(s) can you make that include 4?

74. The sequence that includes 21 also includes 45.

75. What sequence(s) can you make that include 60?

76. The sequence that includes 17 also includes 65.

81. What are the possible common differences?

84. What are the possible common differences?

85. What are the possible common differences?

86. What is the greatest possible common difference? The smallest?

87. What do the third and fourth terms tell you about the common difference?

101. What numbers must fill the empty squares in the left column?

102. Start with the bottom row and the left column. What is the common difference of each? What number can fill the square above the 16?

103. What numbers are missing in the top row? The bottom row? How can we arrange these to make each column an arithmetic sequence?

104. What numbers are missing in the top row? The bottom row? How can we arrange these to make each column an arithmetic sequence?

113. What numbers could you add to the 997th triangular number to get the 1,002nd triangular number?

114. What expression gives the kth triangular number for any number k?

132. Do you see any triangular number patterns in the gumballs?

133. How many rows and columns are there in the kth figure?

134. How many black gumballs are in the nth figure? How many gray gumballs?

138. If we call the second term in Grogg's sequence a, what expression could we write for the third term in Grogg's sequence?

141. How could your answers from the previous two problems help you answer this one?

142. To reach each step, Hoppy can hop to the step directly below it and make a 1-step hop, or he can hop to the step that is two below it and make a 2-step hop.

143. Start small. How many arrangements of 1 domino are possible? 2? 3? 4? How can you use these shorter arrangements to count longer ones?

149. List the first nine terms. Notice a pattern?

152. List the first nine terms. Notice a pattern?

153. You don't have to pick the first terms first.

154. What is true about the digit sum of any multiple of 9?

158. The terms spiral around to form squares. When the first 100 terms are written, where will the 100 be?

160. What number is displayed after one press? Two presses? Continue this sequence and look for a pattern.

161. Consider the first few pours and look for a pattern. Will this pattern continue?

164. How many of the first 50 positive integers are squares?

165. What is the smallest number that has all of its positive multiples included in this sequence?

166. What is the smallest integer that can appear in this sequence?

167. How could you find $y+z$?

168. How many toothpicks are added to each figure to make the next one?

169. What expression could you use to find the kth term of each sequence?

14. What could we multiply both quantities in the ratio by to give a ratio of two integers?

15. What could we multiply both quantities in the ratio by to give a ratio of two integers?

26. We can split Allison's green paint mixture into 5 equal parts, with some blue and some yellow. If there is $\frac{1}{4}$ of a pint of yellow paint, how much paint is in each part?

27. How many miles did Peter travel by train?

28. Which color of bead will Anna run out of first?

33. How many stingravens are there in the exhibit?

34. What fraction of the perimeter comes from the sides that are the rectangle's width?

42. How could you write the cornstarch-to-water ratio as a ratio of integers?

78. What fraction of 100 is a?

79. What fraction of the total number of goals scored by Ben and Alfie is 16?

86. **a.** If Griffinburg and North Pegasus are actually m miles apart, what equation can we write?
b. If Unicorpia and West Mermaid are m inches apart on the map, what equation can we write?

90. What fraction of all the trees are apple trees? Fuji apple trees? Gala apple trees? Peach trees?

94. What fraction of all the animals are green frogs? Yellow frogs? Toads?

95. What fraction are a, b, and c of 70?

96. Using the given ratio, we can label the length, width, and height of the rectangle $2x$, $5x$, and $8x$.

109. What numbers fill the triangles that touch the 3:5:8 square?

110. What numbers fill the triangles that touch the 2:3:5 square?

131. How many minutes does it take Urlick to paddle 1 mile?

135. **b.** Use your answer to part a. How many houses can 9 paint-bots paint in 3 hours?

136. **b.** Use your answer to part a. How many days will it take for 1 carpenter to build 8 sheds? How many days will it take 3 carpenters to build 8 sheds?

137. **a.** How long will it take 24 lumberjacks to chop 5 logs?
b. What fraction is 6 logs of 9 logs? What fraction is 72 minutes of 40 minutes?

142. At 2 meters per second, how many meters can you jog in a minute?

175. How many grams of sugar are in the whole box? How many grams of sugar are in one ounce?

176. What conversion factors would you use to convert from days to weeks, centimeters to inches, and inches to feet?

178. How many of the 135 fruits does Jeremy have?

181. Using the given ratio, we can label the heights of Jorble and Yorble $5x$ and $9x$. What is the difference in their heights?

182. If the height of the rectangle is $2x$, what is its width?

183. If Alyssa bikes 5 mph and runs 2 mph, what is the ratio of the time it takes for her to bike to school to the time it takes for her to run to school?

184. Since the number of red cars doesn't change, how could you rewrite the ratios using the same number of red cars in both ratios?

185. What fraction of the seats are filled? Empty? Filled by teachers?

186. How long does it take the skateboards to collide?

20. What is the smallest decimal that rounds to 0.4 when rounded to the nearest tenth?

46. What is the smallest integer result you could get?

47. What is the smallest integer result you could get?

80. What is 0.A+0.B?

57. How are multiplication and division related?

93. How many digits are to the right of the decimal point if you write the trailing zeros in the product?

94. How many digits are to the right of the decimal point in $(0.3)^{15}$? In $(0.07)^{15}$?

95. How many zeros are at the end of the product $6^{15} \times 5^{15}$?

104. How many trailing zeros are in the product $0.\square\square\square \times 0.\square$ before we remove them to get 0.7?

105. How many trailing zeros are in the product $1.\square \times 0.\square\square\square$ before we remove them to get 0.04?

112. What is 300×0.002?

113. What is 400×0.0025?

118. Try writing these decimals as fractions.

119. What could you do with the rectangles to make this computation easier?

153. How could we write this fraction with a denominator that is a power of 10?

163. What is the repeating pattern of digits?

164. What is the repeating pattern of digits?

171. We know that $\frac{1}{90} = \frac{1}{10} \times \frac{1}{9}$, $\frac{1}{45} = 2 \times \frac{1}{90}$, and $\frac{1}{15} = 6 \times \frac{1}{90}$.

175. Multiplying by $\frac{1}{111}$ is the same as dividing by 111.
Then, we have $\frac{1}{999} = \frac{1}{333} \times \frac{1}{3} = \frac{1}{333} \div 3$.

180. How could your answer from the previous problem help here?

181. We know that $\frac{1}{90} = \frac{1}{10} \times \frac{1}{9}$. Since we know how to write any number of ninths as a decimal, we can write any number of 90ths as a decimal.

195. How many of the three empty squares in the second row contain a digit? What digit(s)?

196. How could you fill in digits to complete the path of the $\frac{1}{4}$ that begins at the bottom of the grid?

199. How could you complete the path for $\frac{5}{9}$ without blocking the path for $\frac{7}{25}$? Then, how could you complete the path for $\frac{17}{40}$?

200. What digits must be placed in the loop that is shared by the paths of $\frac{6}{11}$, $\frac{5}{11}$, and $\frac{17}{11}$? How can these digits be placed?

201. For a puzzle like this, knowing where numbers *cannot* go is really important. Begin by X'ing any location on the outside of the grid whose path does not include exactly one decimal point. As you work through this enormous Frac-Turn, be sure to place X's in squares that cannot contain digits.

215. What is the smallest possible product that has exactly three digits after the decimal point (not including trailing zeros)?

218. What is always true of the denominator of a fraction that can be expressed as a terminating decimal? (It may help to review your answer to Problem 166.)

219. How could you express $0.0\overline{3}$ and $0.\overline{2}$ as fractions?

220. What is $0.3333... = 0.\overline{3}$ as a fraction?

SOLUTIONS
Chapters 7-9

SEQUENCES

What Comes Next?   7-8

1. The pattern begins with an equilateral triangle. Then, each figure in the pattern is a regular polygon with one more side than the figure that came before it.

The last figure we see in the pattern is a regular hexagon. So, the next figure is the regular heptagon.

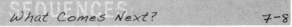

2. The dot's position alternates between the top and bottom of the square.

So, in the next figure the dot will be at the top of the square.

The black triangle moves clockwise from corner to corner.

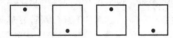

So, in the next figure the black triangle will be in the upper-left corner.

We circle the only answer choice with the correct placement of the dot and the black triangle.

3. The circle's position moves *counterclockwise* from vertex to vertex in the triangle.

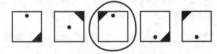

So, in the next figure the circle will be at the top vertex.

The tick marks move *clockwise* around the sides of the triangle, with each figure having one more tick mark than the one before it.

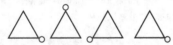

So, in the next figure there will be five tick marks on the right side of the triangle.

We circle the only answer choice with the correct placement of the circle and tick marks.

4. In each figure, a path is drawn from the lower-left vertex of the square to one of the vertices on the right side. The number of segments in the path increases by 1 from each figure to the next.

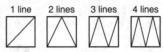

In the next figure, a path of 5 segments will connect the lower-left vertex with the upper-right vertex.

5. The small square moves *clockwise* from corner to corner within the larger square.

So, in the next figure the small square will be in the upper-right corner of the larger square.

The segment within the smaller square rotates 45 degrees *counterclockwise* from one figure to the next.

So, in the next figure the segment will be rotated so that it connects the lower-left and upper-right vertices of the smaller square:

We circle the only answer choice with the correct placement of the smaller square and the line within it.

6. The black triangle moves counterclockwise from point to point within the star.

So, in the next figure the black triangle will be in the rightmost point of the star.

The dot alternates between the right and left sides of the black triangle.

So, in the next figure the dot will be on the right side of the black triangle.

We circle the only answer choice with the correct placement of the black triangle and dot.

7. Each figure is a copy of the figure before it, with one extra diagonal drawn. However, it's difficult to see the pattern the diagonals make as they are being added.

To make the pattern easier to visualize, we label the ends of the diagonal in the first figure A and B, as shown to the right.

In the second figure, we have the diagonal from the first figure, plus a new diagonal. We can think of this new diagonal as a rotation of the original diagonal, as shown below.

The A and B ends each rotate two vertices clockwise. If we continue this pattern, then we get the first four figures that are given.

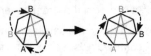

So, to get the next figure we rotate the A and B ends of the newest diagonal two vertices clockwise:

We circle the correct answer choice as shown below.

8. The letters alternate between capitalized and lowercase, and every other letter in the ordered alphabet is skipped.

So, the letter after the lowercase 'k' will be capitalized, and it is two letters after 'k' in the alphabet. This gives the capital letter **M**.

9. Each letter in this list is the first letter of a day of the week, starting with Sunday and moving sequentially through Friday. The day after Friday is Saturday, so the next letter is **S**.

10. Moving from A to E, we skip the letters B-C-D. Then, moving from E to I, we skip the letters F-G-H. Then, moving from I to M, we skip the letters J-K-L.

To get from one letter in the list to the next, we skip the three letters between them (after Y, we skip Z, then loop back through the alphabet and skip A and B.)

So, to find the letter after G, we skip H-I-J to get **K**.

11. The letters in this list are in reverse alphabetical order. Every second letter is also upside down. Since T comes before U, the letter after the upside-down U is a right-side-up **T**.

12. The letters in this list are the first letters of the positive integers when spelled:

One, Two, Three, Four, Five, Six, Seven, Eight

So, the next letter in the pattern is **N** for Nine.

13. The letters shown are in alphabetical order, with some letters missing.

Moving from A to B, we do not skip any letters.
Moving from B to D, we skip the letter C.

Moving from D to G, we skip E-F.
Moving from G to K, we skip H-I-J.
Moving from K to P, we skip L-M-N-O.

Each time we skip a group of letters, we skip one more letter than the previous skip. After P, we skip Q-R-S-T-U. So, the next letter is **V**.

14. The letters in this list are the first letters of the planets in our solar system! Starting with the planet closest to the sun and moving outward, we have

Mercury, Venus, Earth, Mars, Jupiter, Saturn, Uranus.

The next planet is Neptune, so the next letter in the pattern is **N**.

15. Reading from right to left, these letters spell

E A S T A C A D E M Y

These are the letters in "Beast Academy," but with the B missing! So, the next letter is **B**.

16. The letters are arranged into an "upper row" and a "lower row." The letters in each row appear in alphabetical order.

All of the letters in the upper row are written without curves, and all of the letters in the lower row have at least one curve.

The next letter in the pattern belongs in the lower row, so it is the first capital letter after J that has a curve. After J, the letters K, L, M, and N are written without curves, but **O** is curved.

17. If we say the letters in this list out loud, they all rhyme!

"Bee, Cee, Dee, Ee, Gee, Pee, Tee, ..."

These letters are also in alphabetical order. So, the next letter in the list is the first letter after T that rhymes with the letters in the list: **V**.

18. To get from 1 to 2, we add 1.
To get from 2 to 4, we add 2.
To get from 4 to 7, we add 3.
To get from 7 to 11, we add 4.
To get from 11 to 16, we add 5.

The number we add to get from one term in the sequence to the next increases by 1 each time. So, the number after 16 is $16+6=22$, followed by $22+7=29$, followed by $29+8=37$.

$$1, 2, 4, 7, 11, 16, \mathbf{22}, \mathbf{29}, \mathbf{37}$$

19. To get from 46 to 40, we subtract 6.
To get from 40 to 32, we subtract 8.
To get from 32 to 22, we subtract 10.
To get from 22 to 10, we subtract 12.

The number we subtract to get from one term in the sequence to the next increases by 2 each time. So, the number after 10 is $10-14=-4$, followed by $-4-16=-20$, followed by $-20-18=-38$.

$$46, 40, 32, 22, 10, \textbf{-4, -20, -38}$$

20. Ignoring signs, each number after 640 is half the number that came before it. Then, every other term is negative. So, the next three terms are shown below.

$$640, -320, 160, -80, 40, \textbf{-20, 10, -5}$$

We can also find each next term in the sequence by dividing the previous term by -2.

21. The numbers in this sequence alternate between getting bigger and getting smaller.

Each time a number gets bigger, it is double the number that came before it. For example, we double 10 to get 20, we double 14 to get 28, and we double 22 to get 44.

$$\overset{\times 2}{\frown}\ \overset{\times 2}{\frown}\ \overset{\times 2}{\frown}$$
$$10,\ 20,\ 14,\ 28,\ 22,\ 44,\ 38,\ \_\_,\ \_\_,\ \_\_$$

Each time a number gets smaller, we subtract 6 from the number that came before it. For example, we subtract 6 from 20 to get 14, we subtract 6 from 28 to get 22, and we subtract 6 from 44 to get 38.

$$\overset{-6}{\frown}\ \overset{-6}{\frown}\ \overset{-6}{\frown}$$
$$10,\ 20,\ 14,\ 28,\ 22,\ 44,\ 38,\ \_\_,\ \_\_,\ \_\_$$

Continuing this alternating pattern, the next three numbers in the sequence are $38\cdot 2=76$, then $76-6=70$, then $70\cdot 2=140$.

$$\overset{\times 2}{\frown}\overset{-6}{\frown}\overset{\times 2}{\frown}\overset{-6}{\frown}\overset{\times 2}{\frown}\overset{-6}{\frown}\overset{\times 2}{\frown}\overset{-6}{\frown}\overset{\times 2}{\frown}$$
$$10,\ 20,\ 14,\ 28,\ 22,\ 44,\ 38,\ \textbf{76, 70, 140}$$

22. The numbers in this sequence are increasing, which suggests that addition or multiplication may be involved.

We consider a pattern involving addition:

$$\overset{+1}{\frown}\ \overset{+4}{\frown}\ \overset{+18}{\frown}\ \overset{+96}{\frown}$$
$$1,\ 2,\ 6,\ 24,\ 120,\ \_\_,\ \_\_,\ \_\_$$

There is no obvious pattern that uses addition to get from one term to the next. So, we consider multiplication.

$$\overset{\times 2}{\frown}\ \overset{\times 3}{\frown}\ \overset{\times 4}{\frown}\ \overset{\times 5}{\frown}$$
$$1,\ 2,\ 6,\ 24,\ 120,\ \_\_,\ \_\_,\ \_\_$$

Here, we see that the number we multiply by to get from one term to the next increases by 1 each time. So, the next number is $120\cdot 6=720$, then $720\cdot 7=5040$, then $5040\cdot 8=40320$.

$$\overset{\times 2}{\frown}\ \overset{\times 3}{\frown}\ \overset{\times 4}{\frown}\ \overset{\times 5}{\frown}\ \overset{\times 6}{\frown}\ \overset{\times 7}{\frown}\ \overset{\times 8}{\frown}$$
$$1,\ 2,\ 6,\ 24,\ 120,\ \textbf{720, 5040, 40320}$$

You may have noticed that the numbers in this sequence are the factorials: 1!, 2!, 3!, 4!, 5!, 6!, 7!, 8!.

23. The terms in this sequence alternate between fractions and integers. We notice that each of the fraction terms has denominator 2 and a numerator that is a perfect square. However, it is difficult to see a pattern among the terms that are integers: 2, 8, and 18.

So, we try writing the integer terms as equivalent fractions with denominator 2. This gives the following sequence:

$$\frac{1}{2},\ \frac{4}{2},\ \frac{9}{2},\ \frac{16}{2},\ \frac{25}{2},\ \frac{36}{2},\ \_\_,\ \_\_,\ \_\_.$$

The numerators of the terms are the perfect squares! The fractions with even numerators are simplified in the original sequence.

Continuing this pattern, the next three fractions are $\frac{49}{2},\ \frac{64}{2}$, and $\frac{81}{2}$. Simplifying $\frac{64}{2}=32$, we have

$$\frac{1}{2},\ 2,\ \frac{9}{2},\ 8,\ \frac{25}{2},\ 18,\ \frac{49}{2},\ \textbf{32},\ \frac{81}{2}.$$

24. We look for a relationship between each term and its position in the sequence.

The 1st term in the sequence is $1+18=19$.
The 2nd term in the sequence is $2+18=20$.
The 3rd term in the sequence is $3+18=21$, and so on.

Each term is 18 greater than its position number. So, the 15th term is $15+18=\textbf{33}$, and the 50th term is $50+18=\textbf{68}$.

25. Each term in this sequence is a number raised to the third power.

The 1st term is $1^3=1$, the 2nd term is $2^3=8$, the 3rd term is $3^3=27$, the 4th term is $4^3=64$, the 5th term is $5^3=125$, and so on.

Each term is equal to its position number raised to the third power. So, the 9th term is $9^3=\textbf{729}$, and the 20th term is $20^3=\textbf{8,000}$.

26. The terms in this sequence alternate between positive and negative multiples of 3.

The 1st term is $3\cdot 1=3$, the 2nd term is $-(3\cdot 2)=-6$, the 3rd term is $3\cdot 3=9$, the 4th term is $-(3\cdot 4)=-12$, and so on.

Each term whose position number is *odd* is 3 times its position in the sequence. Each term whose position number is *even* is the *opposite* of 3 times its position in the sequence.

So, the 25th term in the sequence is $3\cdot 25=\textbf{75}$, and the 100th term in the sequence is $-(3\cdot 100)=\textbf{-300}$.

27. The terms in this sequence alternate between 0 and some multiple of 7. The terms in odd-numbered positions are equal to 0: the 1st term, 3rd term, 5th term, and so on. So, the 99th term is **0**.

Among the terms whose position number is even, we see the following pattern:

The 2nd term is $1\cdot 7=7$.
The 4th term is $2\cdot 7=14$.
The 6th term is $3\cdot 7=21$.

Each of these terms is equal to the product of half its position number and 7. So, the 40th term in the sequence is $(40\div 2)\cdot 7=20\cdot 7=\textbf{140}$.

28. The terms in this sequence form a pattern that repeats every four terms, as shown below:

$$\underbrace{1,\ 11,\ 111,\ 11,}\ \underbrace{1,\ 11,\ 111,\ 11,}\ \underbrace{1,\ \ldots}$$

Within this pattern, every second term is 11. So, any term with an even position number is 11. Therefore, the 80th term in the sequence is **11**.

Since 51 is odd, the 51st term is either 1 or 111.

The terms equal to 1 are the 1st, 5th, 9th, 13th, and so on. These are the terms whose position number is 1 more than a multiple of 4.

The terms equal to 111 are the 3rd, 7th, 11th, 15th, and so on. These are the terms whose position number is 3 more than a multiple of 4.

Since $51 = 48+3$ is three more than a multiple of 4, the 51st term is **111**.

29. This is the sequence of the positive even numbers. Each term is equal to twice its position number.

Position: 1st $\quad$ 2nd $\quad$ 3rd $\quad$ 4th $\quad$ 5th $\quad\cdots$
$$\quad\quad\ \searrow{\scriptstyle 2\cdot1}\ \searrow{\scriptstyle 2\cdot2}\ \searrow{\scriptstyle 2\cdot3}\ \searrow{\scriptstyle 2\cdot4}\ \searrow{\scriptstyle 2\cdot5}$$
Term: 2 $\quad$ 4 $\quad$ 6 $\quad$ 8 $\quad$ 10 $\quad\cdots$

So, the 25th term is $2\cdot25 = 50$, the 50th term is $2\cdot50 = 100$, and the nth term is $2n$.

Position:	1st	2nd	3rd	4th	5th	$\cdots$	25th	$\cdots$	50th	$\cdots$	nth
Term:	2	4	6	8	10	$\cdots$	**50**	$\cdots$	**100**	$\cdots$	**$2n$**

30. This is the sequence of odd numbers, beginning with 1. We notice that each term is one less than its corresponding term in the sequence of even integers from the previous problem. So, each term is one less than twice its position number.

Position: 1st $\quad$ 2nd $\quad$ 3rd $\quad$ 4th $\quad$ 5th $\quad\cdots$
$$\quad\ \searrow{\scriptstyle 2\cdot1-1}\ \searrow{\scriptstyle 2\cdot2-1}\ \searrow{\scriptstyle 2\cdot3-1}\ \searrow{\scriptstyle 2\cdot4-1}\ \searrow{\scriptstyle 2\cdot5-1}$$
Term: 1 $\quad$ 3 $\quad$ 5 $\quad$ 7 $\quad$ 9 $\quad\cdots$

So, the 30th term is $2\cdot30-1 = 59$, the 75th term is $2\cdot75-1 = 149$, and the nth term is $2n-1$.

Position:	1st	2nd	3rd	4th	5th	$\cdots$	30th	$\cdots$	75th	$\cdots$	nth
Term:	1	3	5	7	9	$\cdots$	**59**	$\cdots$	**149**	$\cdots$	**$2n-1$**

31. Each term in this sequence is equal to 2 raised to the power of its position number.

Position: 1st $\quad$ 2nd $\quad$ 3rd $\quad$ 4th $\quad$ 5th $\quad\cdots$
$$\quad\quad\ \searrow{\scriptstyle 2^1}\ \searrow{\scriptstyle 2^2}\ \searrow{\scriptstyle 2^3}\ \searrow{\scriptstyle 2^4}\ \searrow{\scriptstyle 2^5}$$
Term: 2 $\quad$ 4 $\quad$ 8 $\quad$ 16 $\quad$ 32 $\quad\cdots$

So, the 6th term is $2^6 = 64$, the 10th term is $2^{10} = 1{,}024$, and the nth term is 2^n.

Position:	1st	2nd	3rd	4th	5th	$\cdots$	6th	$\cdots$	10th	$\cdots$	nth
Term:	2	4	8	16	32	$\cdots$	**64**	$\cdots$	**1,024**	$\cdots$	**2^n**

32. This is the sequence of perfect squares. Each term is equal to the square of its position number.

Position: 1st $\quad$ 2nd $\quad$ 3rd $\quad$ 4th $\quad$ 5th $\quad\cdots$
$$\quad\quad\ \searrow{\scriptstyle 1^2}\ \searrow{\scriptstyle 2^2}\ \searrow{\scriptstyle 3^2}\ \searrow{\scriptstyle 4^2}\ \searrow{\scriptstyle 5^2}$$
Term: 1 $\quad$ 4 $\quad$ 9 $\quad$ 16 $\quad$ 25 $\quad\cdots$

So, the 10th term is $10^2 = 100$, the 25th term is $25^2 = 625$, and the nth term is n^2.

Position:	1st	2nd	3rd	4th	5th	$\cdots$	10th	$\cdots$	25th	$\cdots$	nth
Term:	1	4	9	16	25	$\cdots$	**100**	$\cdots$	**625**	$\cdots$	**n^2**

33. The terms in this sequence alternate between -1 and 1. We know that odd powers of -1 are equal to -1, and even powers of -1 are equal to 1. So, we can write each term as -1 raised to the power of its position number.

Position: 1st $\quad$ 2nd $\quad$ 3rd $\quad$ 4th $\quad$ 5th $\quad\cdots$
$$\quad\ \searrow{\scriptstyle (-1)^1}\ \searrow{\scriptstyle (-1)^2}\ \searrow{\scriptstyle (-1)^3}\ \searrow{\scriptstyle (-1)^4}\ \searrow{\scriptstyle (-1)^5}$$
Term: -1 $\quad$ 1 $\quad$ -1 $\quad$ 1 $\quad$ -1 $\quad\cdots$

So, the 15th term is $(-1)^{15} = -1$, the 40th term is $(-1)^{40} = 1$, and the nth term is $(-1)^n$.

Position:	1st	2nd	3rd	4th	5th	$\cdots$	15th	$\cdots$	40th	$\cdots$	nth
Term:	-1	1	-1	1	-1	$\cdots$	**-1**	$\cdots$	**1**	$\cdots$	**$(-1)^n$**

34. We begin by writing 2 as $\frac{2}{1}$ and 1 as $\frac{1}{1}$ so that every term in the sequence is a fraction.

$$\frac{2}{1},\ \frac{1}{1},\ \frac{2}{3},\ \frac{1}{2},\ \frac{2}{5},\ \frac{1}{3},\ \frac{2}{7},\ \frac{1}{4},\ \cdots$$

Then, we notice that the numerators in this sequence alternate between 2 and 1. We could consider the terms with even position numbers separately from the terms with odd position numbers. However, doing so makes it difficult to write an expression for the nth term, whose position number could be even or odd.

We can write each fraction with numerator 1 as an equivalent fraction with numerator 2. This way, all fractions have the same numerator, as shown below.

$$\frac{2}{1},\ \frac{2}{2},\ \frac{2}{3},\ \frac{2}{4},\ \frac{2}{5},\ \frac{2}{6},\ \frac{2}{7},\ \frac{2}{8},\ \cdots$$

Now, each term is a fraction with a numerator of 2 and a denominator equal to that term's position number. So, the 20th term is $\frac{2}{20} = \frac{1}{10}$, the 35th term is $\frac{2}{35}$, and the nth term is $\frac{2}{n}$.

SEQUENCES
Arithmetic Sequences, Part 1 $\qquad$ 12-15

35. We add 9 to get from each term to the next.

$$\overset{+9}{\frown}\ \overset{+9}{\frown}\ \overset{+9}{\frown}\ \overset{+9}{\frown}$$
$$7,\ 16,\ 25,\ 34,\ 43,\ \ldots$$

So, the common difference is **9**.

36. We add 8 to get from each term to the next.

$$\overset{+8}{\frown}\ \overset{+8}{\frown}\ \overset{+8}{\frown}\ \overset{+8}{\frown}$$
$$-33,\ -25,\ -17,\ -9,\ -1,\ \ldots$$

So, the common difference is **8**.

37. We add -3 to get from each term to the next.

$$\overset{+(-3)\ \ +(-3)\ \ +(-3)\ \ +(-3)}{29,\ 26,\ 23,\ 20,\ 17,\ ...}$$

So, the common difference is **-3**.

38. To get from 21 to 35, we add the common difference twice.

$$\underline{\ \ },\ 21,\ \underline{\ \ },\ 35,\ \underline{\ \ },\ ...$$

Adding the common difference twice adds a total of $35-21=14$. So, the common difference is $14\div 2=$ **7**.

39. To get from 74 to 41, we add the common difference 3 times.

$$74,\ \underline{\ \ },\ \underline{\ \ },\ 41,\ \underline{\ \ },\ ...$$

Adding the common difference 3 times adds a total of $41-74=-33$. So, the common difference is $-33\div 3=$ **-11**.

40. To get from 30 to $32\frac{1}{2}$, we add the common difference 5 times.

$$30,\ \underline{\ \ },\ \underline{\ \ },\ \underline{\ \ },\ \underline{\ \ },\ 32\tfrac{1}{2},\ ...$$

Adding the common difference 5 times adds a total of $32\frac{1}{2}-30=2\frac{1}{2}=\frac{5}{2}$.

So, the common difference is $\frac{5}{2}\div 5=\frac{5}{2}\cdot\frac{1}{5}=\frac{1}{2}$.

41. To get from 19 to 64, we add the common difference 3 times. This adds a total of $64-19=45$. So, the common difference is $45\div 3=15$.

We use the common difference to fill in the blanks as shown.

$$\overset{+15\ \ +15\ \ +15\ \ +15\ \ +15\ \ +15}{19,\ \underline{\textbf{34}},\ \underline{\textbf{49}},\ 64,\ \underline{\textbf{79}},\ \underline{\textbf{94}},\ 109}$$

42. To get from 98 to 112, we add the common difference twice. This adds a total of $112-98=14$. So, the common difference is $14\div 2=7$.

We use the common difference to fill in the blanks as shown.

$$\overset{+7\ \ +7\ \ +7\ \ +7\ \ +7\ \ +7}{98,\ \underline{\textbf{105}},\ 112,\ \underline{\textbf{119}},\ \underline{\textbf{126}},\ \underline{\textbf{133}},\ 140}$$

43. To get from 32 to $30\frac{1}{2}$, we subtract $1\frac{1}{2}$ (or add $-1\frac{1}{2}$). So, the term *after* $30\frac{1}{2}$ is $30\frac{1}{2}-1\frac{1}{2}=29$.

Working backwards, we can find the terms that come *before* 32 by continually *adding* $1\frac{1}{2}$, as shown below.

$$\overset{+1\frac{1}{2}\ +1\frac{1}{2}\ +1\frac{1}{2}\ +1\frac{1}{2}\qquad\ -1\frac{1}{2}}{\underline{\textbf{38}},\ \underline{\textbf{36}\tfrac{1}{2}},\ \underline{\textbf{35}},\ \underline{\textbf{33}\tfrac{1}{2}},\ 32,\ 30\tfrac{1}{2},\ \underline{\textbf{29}}}$$

44. To get from 10 to 22 we add the common difference 5 times. This adds a total of $22-10=12$. So, the common difference is $12\div 5=\frac{12}{5}=2\frac{2}{5}$.

We use the common difference to fill in the blanks as shown.

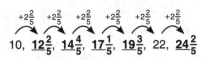
$$\overset{+2\frac{2}{5}\ +2\frac{2}{5}\ +2\frac{2}{5}\ +2\frac{2}{5}\ +2\frac{2}{5}\ +2\frac{2}{5}}{10,\ \underline{\textbf{12}\tfrac{2}{5}},\ \underline{\textbf{14}\tfrac{4}{5}},\ \underline{\textbf{17}\tfrac{1}{5}},\ \underline{\textbf{19}\tfrac{3}{5}},\ 22,\ \underline{\textbf{24}\tfrac{2}{5}}}$$

45. To get from the 1st term to the 10th term in an arithmetic sequence, we add the common difference $10-1=9$ times. This adds a total of $115-25=90$. So, the common difference is $90\div 9=$ **10**.

46. To get from the 23rd term to the 25th term, we add the common difference $25-23=2$ times. This adds a total of $\frac{1}{2}-\frac{1}{3}=\frac{1}{6}$.

So, the common difference is $\frac{1}{6}\div 2$. Since dividing by a number is the same as multiplying by that number's reciprocal, we have $\frac{1}{6}\div 2=\frac{1}{6}\cdot\frac{1}{2}=\frac{1}{12}$.

47. To get from the 10th term to the 30th term, we add the common difference $30-10=20$ times. This adds a total of $68-4=64$. So, the common difference is $\frac{64}{20}=\frac{16}{5}$.

We get the 20th term by adding the common difference 10 times to the 10th term. The 10th term is 4, so the 20th term is $4+10\cdot\frac{16}{5}=4+32=$ **36**.

— *or* —

We add the common difference 10 times to get from the 10th to the 20th term, and 10 times to get from the 20th term to the 30th term. So, the 20th term is halfway between the 10th and 30th terms. The number halfway between 4 and 68 is their average: $\frac{4+68}{2}=\frac{72}{2}=$ **36**.

48. If we let d represent the common difference of the sequence, then the five terms in the sequence are

$$40,\ 40+d,\ 40+2d,\ 40+3d,\ 40+4d.$$

The sum of these terms is $40(5)+(d+2d+3d+4d)$, which simplifies to $200+10d$.

We are told that the sum of all five terms is 80. This gives the equation $200+10d=80$. Subtracting 200 from both sides, we have $10d=-120$. Dividing both sides by 10 gives $d=-12$.

So, the common difference of the sequence is **-12**.

— *or* —

Since the difference between consecutive terms in an arithmetic sequence is always the same, the five terms in this sequence balance around the middle term.

$$\overset{-d\quad -d\ \ \vdots\ \ +d\quad +d}{\underset{\text{1st}\quad \text{2nd}\quad \text{3rd}\quad \text{4th}\quad \text{5th}}{40\quad ?\quad ?\quad ?\quad ?}}$$

So, the 3rd term is equal to the average of the terms. The sum of the five terms in the sequence is 80, so the average is $\frac{80}{5}=16$. Therefore, the 3rd term is 16.

$$\underset{\text{1st}\quad \text{2nd}\quad \text{3rd}\quad \text{4th}\quad \text{5th}}{40\quad ?\quad 16\quad ?\quad ?}$$

We add the common difference twice to get from the 1st term of 40 to the 3rd term of 16. So, the common difference is $\frac{16-40}{2}=\frac{-24}{2}=$ **-12**.

49. The first term of the sequence is 15 and the common difference is 4.

$$\overset{+4}{\frown}\ \overset{+4}{\frown}\ \overset{+4}{\frown}\ \overset{+4}{\frown}$$
$$15,\ 19,\ 23,\ 27,\ 31,\ \dots$$

To get to the 2nd term, we add 1 four to 15.
To get to the 3rd term, we add 2 fours to 15.
To get to the 4th term, we add 3 fours to 15.

To get to the 10th term, we add 9 fours to 15.
So, the 10th term is $15+9(4)=15+36=\textbf{51}$.

50. The first term of the sequence is -11 and the common difference is 5.

$$\overset{+5}{\frown}\ \overset{+5}{\frown}\ \overset{+5}{\frown}\ \overset{+5}{\frown}$$
$$-11,\ -6,\ -1,\ 4,\ 9,\ \dots$$

To get to the 2nd term, we add 1 five to -11.
To get to the 3rd term, we add 2 fives to -11.
To get to the 4th term, we add 3 fives to -11.

To get to the 40th term, we add 39 fives to -11.
So, the 40th term is $-11+39(5)=-11+195=\textbf{184}$.

51. The first term of the sequence is 5 and the common difference is -7.

$$\overset{+(-7)}{\frown}\ \overset{+(-7)}{\frown}\ \overset{+(-7)}{\frown}\ \overset{+(-7)}{\frown}$$
$$5,\ -2,\ -9,\ -16,\ -23,\ \dots$$

To get to the 2nd term, we add one -7 to 5.
To get to the 3rd term, we add two -7's to 5.
To get to the 4th term, we add three -7's to 5.

To get to the 15th term, we add fourteen -7's to 5.
So, the 15th term is $5+14(-7)=5+(-98)=\textbf{-93}$.

52. The first term of the sequence is -29 and the common difference is 10.

$$\overset{+10}{\frown}\ \overset{+10}{\frown}\ \overset{+10}{\frown}\ \overset{+10}{\frown}$$
$$-29,\ -19,\ -9,\ 1,\ 11,\ \dots$$

To get to the 100th term, we add 99 tens to -29.
So, the 100th term is $-29+99(10)=-29+990=\textbf{961}$.

53. To get to the 13th term, we add the common difference to the first term 12 times. The first term is 9, and the common difference is 8. So, the 13th term is $9+12(8)=9+96=\textbf{105}$.

54. To go from the 1st term to the 100th term, we *add* the common difference 99 times. So, to go from the 100th term to the 1st term, we *subtract* the common difference 99 times.

The 100th term is 40, and the common difference is $\frac{1}{3}$. So, the first term is $40-99\cdot\frac{1}{3}=40-33=\textbf{7}$.

55. To go from the 12th term to the 15th term, we add the common difference 3 times. The difference between the 12th and 15th terms is $106-85=21$. So, the common difference is $\frac{106-85}{3}=\frac{21}{3}=7$.

Then, we find the first term of the sequence by subtracting the common difference from the 12th term 11 times. So, the first term is $85-11(7)=85-77=\textbf{8}$.

56. The first term of the sequence is 18 and the common difference is 6.

$$\overset{+6}{\frown}\ \overset{+6}{\frown}\ \overset{+6}{\frown}\ \overset{+6}{\frown}$$
$$18,\ 24,\ 30,\ 36,\ 42,\ \dots$$

To get to the 2nd term, we add 1 six to 18.
To get to the 3rd term, we add 2 sixes to 18.
To get to the 4th term, we add 3 sixes to 18.

To get to the nth term, we add $(n-1)$ sixes to 18.
So, the nth term is $18+(n-1)6$. Distributing the 6 and simplifying gives $18+(n-1)6=18+6n-6=\textbf{6}\boldsymbol{n}\textbf{+12}$.

57. The first term of the sequence is 4 and the common difference is 15.

$$\overset{+15}{\frown}\ \overset{+15}{\frown}\ \overset{+15}{\frown}$$
$$4,\ 19,\ 34,\ 49,\ \dots$$

To get to the 2nd term, we add 1 fifteen to 4.
To get to the 3rd term, we add 2 fifteens to 4.
To get to the 4th term, we add 3 fifteens to 4.

To get to the nth term, we add $(n-1)$ fifteens to 4.
So, the nth term is $4+(n-1)15$. Distributing the 15 and simplifying gives $4+(n-1)15=4+15n-15=\textbf{15}\boldsymbol{n}\textbf{-11}$.

58. The first term of the sequence is -13 and the common difference is 8.

$$\overset{+8}{\frown}\ \overset{+8}{\frown}$$
$$-13,\ -5,\ 3,\ \dots$$

To get to the nth term, we add $(n-1)$ eights to -13.
So, the nth term is $-13+(n-1)8$. Distributing the 8 and simplifying gives $-13+(n-1)8=-13+8n-8=\textbf{8}\boldsymbol{n}\textbf{-21}$.

59. The first term of the sequence is $\frac{9}{4}$ and the common difference is $\frac{5}{2}-\frac{9}{4}=\frac{10}{4}-\frac{9}{4}=\frac{1}{4}$.

$$\overset{+\frac{1}{4}}{\frown}\ \overset{+\frac{1}{4}}{\frown}$$
$$\frac{9}{4},\ \frac{5}{2},\ \frac{11}{4},\ \dots$$

To get to the nth term, we add $(n-1)$ one-fourths to $\frac{9}{4}$.
So, the nth term is $\frac{9}{4}+(n-1)\cdot\frac{1}{4}$. Distributing the $\frac{1}{4}$ and simplifying gives

$$\frac{9}{4}+(n-1)\cdot\frac{1}{4}=\frac{9}{4}+\frac{1}{4}n-\frac{1}{4}$$
$$=\textbf{}\frac{1}{4}\boldsymbol{n}\textbf{+2}.$$

— *or* —

Rewriting $\frac{5}{2}$ as the equivalent fraction $\frac{10}{4}$, we have

$$\frac{9}{4},\ \frac{10}{4},\ \frac{11}{4},\ \dots$$

Each term in the sequence is a fraction with a denominator of 4 and a numerator that is 8 more than its position number in the sequence. For example,

the 1st term is $\frac{1+8}{4}=\frac{9}{4}$,

the 2nd term is $\frac{2+8}{4}=\frac{10}{4}$,

the 3rd term is $\frac{3+8}{4}=\frac{11}{4}$, and so on.

So, the nth term in the sequence is $\frac{n+8}{4}$.

Note that this expression is equivalent to the one we arrived at in the first solution to this problem. We have

$$\frac{1}{4}n+2=\frac{n}{4}+2=\frac{n}{4}+\frac{8}{4}=\frac{n+8}{4}.$$

60. To get to the 20ᵗʰ term of the sequence, we add the common difference to the first term 19 times. The first term is a, and the common difference is 3. So, the 20ᵗʰ term is $a+19(3) = \boldsymbol{a+57}$.

61. To get to the 101ˢᵗ term of the sequence, we add the common difference to the first term 100 times. The first term is 6, and the common difference is d. So, the 101ˢᵗ term is $\boldsymbol{6+100d}$, or $\boldsymbol{100d+6}$.

62. The 1ˢᵗ term of the sequence is 20 and the 2ⁿᵈ term is 32, so the common difference is $32-20 = 12$.

To get the kᵗʰ term in the sequence, we add the common difference to the first term $(k-1)$ times. So, the kᵗʰ term is $20+(k-1)12 = 20+12k-12 = 12k+8$.

We want to know the value of k that makes $12k+8$ equal to 500. This gives the equation

$$12k+8 = 500.$$

Subtracting 8 from both sides, we have $12k = 492$. Dividing both sides by 12 gives $k = \boldsymbol{41}$.

Check: the 41ˢᵗ term is $20+40(12) = 20+480 = 500.$ ✔

SEQUENCES

Sequence Paths   16-17

63. There are five numbers in the grid, and each path must include at least three numbers. So, all five numbers are part of the same path, forming an arithmetic sequence with common difference 7.

The only way to connect all five numbers in an order that forms an arithmetic sequence is shown below.

64. The 21 in the top-left square is the largest number in the grid. So, it must be one end of a path. Also, the path from 21 must pass through 12 or 15 next.

If the path begins 21-12, then the next number is 3. However, the four remaining numbers (7, 9, 15, 17) do not form an arithmetic sequence, and none of them can be part of the 21-12-3 path.

If the path begins 21-15, then the next number is 9. The four remaining numbers (3, 7, 12, 17) do not form an arithmetic sequence. However, 3 can be included at the end of the 21-15-9 path, and the remaining numbers form an arithmetic sequence: 7-12-17.

We connect the numbers in 21-15-9-3 and 7-12-17 as shown.

65. The smallest number in the grid is 3, so it is one end of a path. The only numbers we can draw a path to from 3 are 4, 6, 10, and 12. We consider each possibility.

If the path begins 3-4, then the next number is 5.
If the path begins 3-6, then the next number is 9.
If the path begins 3-10, then the next number is 17.
If the path begins 3-12, then the next number is 21.

Only 3-12-21 has three numbers that appear in the grid. The common difference of 3-12-21 is 9. Since $21+9 = 30$ is not in the grid, there are no other numbers in this path.

The remaining numbers are 4, 6, 8, and 10, which form an arithmetic sequence with common difference 2.

So, we connect 3-12-21 and 4-6-8-10 as shown.

66. We write the numbers in the grid from least to greatest:

12, 18, 24, 26, 30, 34.

Since all six numbers do not form an arithmetic sequence, there must be two 3-term arithmetic sequences.

Since 12 is the smallest number in the grid, it is one end of a path. There is only one arithmetic sequence we can make that begins with 12.

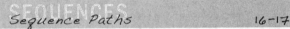

So, one sequence is 12-18-24, and the remaining numbers form the arithmetic sequence 26-30-34.

We connect the numbers in these sequences as shown.

67. We write the numbers in the grid from least to greatest:

56, 57, 61, 62, 66, 67.

Since all six numbers do not form an arithmetic sequence, there must be two 3-term arithmetic sequences.

Since 56 is the smallest number in the grid, it is one end of a path. There is only one arithmetic sequence we can make that begins with 56.

56, 57, 61, 62, 66, 67

So, one sequence is 56-61-66, and the remaining numbers form the arithmetic sequence 57-62-67.

We connect the numbers in these sequences as shown.

68. We write the numbers in the grid from least to greatest:

22, 24, 26, 30, 32, 34, 38, 40.

The only arithmetic sequences we can make that start

with 22 begin with 22-24-26 or 22-26-30 or 22-30-38. It is impossible to draw a path through 22-24-26 or 22-30-38 in the grid. So, one path begins with 22-26-30.

22, 24, 26, 30, 32, 34, 38, 40

Then, the only arithmetic sequence we can make with the smallest remaining number (24) is 24-32-40.

22, 24, 26, 30, 32, 34, 38, 40

Finally, the two remaining numbers (34 and 38) must go at the end of the sequence 22-26-30.

22, 24, 26, 30, 32, 34, 38, 40

We connect the numbers in 22-26-30-34-38 and 24-32-40 as shown.

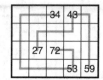

69. We write the numbers in the grid from least to greatest:

27, 34, 43, 53, 59, 72.

The only arithmetic sequence we can make that begins with 27 is 27-43-59. The remaining three numbers form the arithmetic sequence 34-53-72.

We connect the numbers in these sequences as shown.

70. We write the numbers in the grid from least to greatest:

74, 77, 80, 83, 86, 89, 92.

All seven numbers form an arithmetic sequence with common difference 3. However, it is not possible to draw a path through all seven numbers. So, we must have one path with 3 terms and one path with 4 terms. There are only three possibilities:

74, 77, 80, 83, 86, 89, 92

74, 77, 80, 83, 86, 89, 92

74, 77, 80, 83, 86, 89, 92

In the first case, there is no way to connect 74-77-80 without blocking the path for 83-86-89-92. Similarly, in the third case, there is no way to connect 74-80-86-92 without blocking the path for 77-83-89.

So, the two sequences are 74-77-80-83 and 86-89-92. We connect the numbers in these sequences as shown.

71. We write the numbers in order and consider sequences that start with 10. Dashed lines indicate numbers that *might* be included in the sequence.

10, 22, 28, 34, 40, 46, 52, 58

10, 22, 28, 34, 40, 46, 52, 58

10, 22, 28, 34, 40, 46, 52, 58

In the second and third cases shown above, there is no way to make an arithmetic sequence with the five remaining numbers. So, one path begins with 10-22-34, as shown in the first case above.

The only arithmetic sequence we can make with the smallest remaining number (28) is 28-40-52.

The remaining numbers (46 and 58) can only be included at the end of the 10-22-34 sequence.

10, 22, 28, 34, 40, 46, 52, 58

We connect the numbers in 10-22-34-46-58 and 28-40-52 as shown.

72. We write the numbers in the grid from least to greatest:

14, 27, 40, 53, 66, 79, 92.

All seven numbers form an arithmetic sequence with common difference 13. However, it is not possible to draw a path through all seven numbers. So, we must have one path with 3 terms and one path with 4 terms. There are only three possibilities:

14, 27, 40, 53, 66, 79, 92

14, 27, 40, 53, 66, 79, 92

14, 27, 40, 53, 66, 79, 92

In the third case above, if we connect 14-40-66 then there is no way to connect 27 and 53 in the grid.

In each of the first two cases, one path includes 14-27-40 and the other path includes 66-79-92. We must determine which of these two paths includes 53.

We temporarily ignore the 53, and draw both paths. Then, the 53 can only be connected to the end of the 14-27-40 path, as shown.

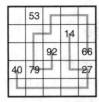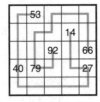

73. The smallest number in the grid is 4. The only arithmetic sequence we can make that begins with 4 is 4-12-20.

4, 5, 9, 12, 13, 15, 17, 20, 25

The six remaining numbers do not form a single arithmetic sequence, so they must form two 3-term arithmetic sequences. There is only one way we can make two 3-term arithmetic sequences, as shown below.

5, 9, 13, 15, 17, 25

We connect the numbers in the sequences 4-12-20, 5-15-25, and 9-13-17 as shown.

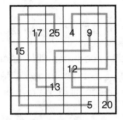

74. We write the numbers in order. These are the numbers we use to skip-count by 6 from 9 to 81, excluding 15:

9, 21, 27, 33, 39, 45, 51, 57, 63, 69, 75, 81.

We consider sequences with smallest term 9:

9-21-33 cannot be connected in order on the grid.

Connecting 9-27-45 forces 81 to be connected to 33 or 75. We cannot make an arithmetic sequence that begins 81-33, and connecting 81-75-69 isolates the 33.

Connecting 9-33-57 as shown below forces 75 to be connected to 63, creating a sequence whose third term is 51. So, if 9-33-57 are on the same path, 75-63-51 must also be on the same path.

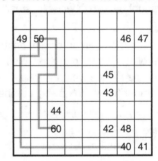

Since 81 cannot be part of a sequence that includes any of the remaining numbers, we must connect 81 to the end of the 9-33-57 path.

Then, since 69 cannot be on either of the existing paths, it must be part of a new path. The only sequence that can be made with the remaining numbers is 69-45-21.

We connect these numbers as shown.

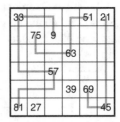

Finally, we can connect 39 and 27 to complete the path that begins 75-63-51.

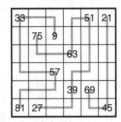

Connecting 9-39-69 leaves a set of numbers that are impossible to group into paths, and connecting 9-45-81 isolates 27. There is no other sequence that includes 9.

So, this is the only solution.

75. The largest number in the grid is 60, the second-largest number is 50, and the smallest number is 40. So, the only arithmetic sequence we can make with 60 is 60-50-40. We connect these numbers as shown below.

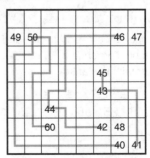

This leaves us with the following numbers:

41, 42, 43, 44, 45, 46, 47, 48, 49.

The sequence that begins with 41 must have a second term that is 45 or less. However, connecting 41 with 42, 44, or 45 will isolate a group of numbers that do not form an arithmetic sequence. So, one sequence begins with 41-43-45.

Then, the only arithmetic sequence we can make with first term 42 begins with 42-44-46.

We connect the numbers in 41-43-45 and 42-44-46 in the only way that does not isolate other numbers in the grid.

Finally, there is no way to connect the remaining three numbers (47, 48, 49) in an order that forms an arithmetic sequence. So, 47 and 49 are included at the end of the 41-43-45 path, and 48 is included at the end of the 42-44-46 path.

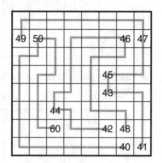

76. We can organize the numbers in this grid into a group of "small numbers" and a group of "big numbers."

Small numbers: 1, 5, 9, 13, 17
Big numbers: 41, 45, 49, 53, 57, 61, 65

Since we cannot connect all five small numbers in the grid without isolating other numbers, there must be a path that begins with a small number and includes a big number.

The greatest possible difference between two small numbers is 17−1=16. The least possible difference between a big and a small number 41−17=24. So, there cannot be a path with two small numbers and a big number.

Therefore, there must be a path that begins with a small number followed by two or more big numbers. The only arithmetic sequence we can make that begins with a small number followed by a big number is 17-41-65.

The remaining small numbers must form the arithmetic sequence 1-5-9-13, and the five remaining big numbers must form the arithmetic sequence 45-49-53-57-61.

We connect the numbers in these sequences in the only way possible, as shown below.

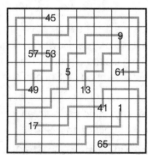

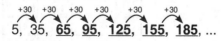

Arithmetic Sequences, Part 2   18–19

77. a. The sequence has 1st term 5 and 2nd term 35, so the common difference is 35−5=30. We use this to fill in the remaining blanks as shown.

$$\overset{+30}{\frown}\ \overset{+30}{\frown}\ \overset{+30}{\frown}\ \overset{+30}{\frown}\ \overset{+30}{\frown}\ \overset{+30}{\frown}$$
5, 35, **65**, **95**, **125**, **155**, **185**, ...

b. The sequence has 1st term 5 and 3rd term 35, so the common difference is $\frac{35-5}{2}=\frac{30}{2}=15$. We use this to fill in the remaining blanks as shown.

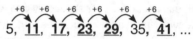
5, **20**, 35, **50**, **65**, **80**, **95**, ...

c. The sequence has 1st term 5 and 4th term 35, so the common difference is $\frac{35-5}{3}=\frac{30}{3}=10$. We use this to fill in the remaining blanks as shown.

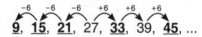
5, **15**, **25**, 35, **45**, **55**, **65**, ...

d. The sequence has 1st term 5 and 6th term 35, so the common difference is $\frac{35-5}{5}=\frac{30}{5}=6$. We use this to fill in the remaining blanks as shown.

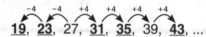
5, **11**, **17**, **23**, **29**, 35, **41**, ...

78. a. We add the common difference twice to get from 27 to 39. So, the common difference is $\frac{39-27}{2}=\frac{12}{2}=6$. We use this to fill in the remaining blanks as shown.

$$\overset{-6}{\curvearrowleft}\ \overset{-6}{\curvearrowleft}\ \overset{-6}{\curvearrowleft}\ \overset{+6}{\frown}\ \overset{+6}{\frown}\ \overset{+6}{\frown}$$
9, **15**, **21**, 27, **33**, 39, **45**, ...

b. We add the common difference 3 times to get from 27 to 39. So, the common difference is $\frac{39-27}{3}=\frac{12}{3}=4$. We use this to fill in the remaining blanks as shown.

$$\overset{-4}{\curvearrowleft}\ \overset{-4}{\curvearrowleft}\ \overset{+4}{\frown}\ \overset{+4}{\frown}\ \overset{+4}{\frown}\ \overset{+4}{\frown}$$
19, **23**, 27, **31**, **35**, 39, **43**, ...

c. We add the common difference 4 times to get from 27 to 39. So, the common difference is $\frac{39-27}{4}=\frac{12}{4}=3$. We use this to fill in the remaining blanks as shown.

$$\overset{-3}{\curvearrowleft}\ \overset{+3}{\frown}\ \overset{+3}{\frown}\ \overset{+3}{\frown}\ \overset{+3}{\frown}\ \overset{+3}{\frown}$$
24, 27, **30**, **33**, **36**, 39, **42**, ...

d. We add the common difference 6 times to get from 27 to 39. So, the common difference is $\frac{39-27}{6}=\frac{12}{6}=2$. We use this to fill in the remaining blanks as shown.

$$\overset{+2}{\frown}\ \overset{+2}{\frown}\ \overset{+2}{\frown}\ \overset{+2}{\frown}\ \overset{+2}{\frown}\ \overset{+2}{\frown}$$
27, **29**, **31**, **33**, **35**, **37**, 39, ...

79. To get from 9 to 33, we add a total of 33−9=24. Since every term is an integer, the common difference must be a factor of 24.

The factors of 24 are 1, 2, 3, 4, 6, 8, 12, and 24. So, there are **8** possible common differences.

80. To get from 15 to 46, we add a total of 46−15=31. Since every term is an integer, the common difference must be a factor of 31.

31 is prime, so its only two factors are 1 and 31. So, there are **2** possible common differences.

81. The difference between 37 and 13 is 37−13=24. Since every term is an integer, the common difference of the sequence is some factor of 24.

We test each possible common difference, stopping if we come across more than one prime between 13 and 37.

1: 13, 14, 15, 16, ⑰, 18, ⑲, ... ✗

2: 13, 15, ⑰, ⑲, ... ✗

3: 13, 16, ⑲, 22, 25, 28, ㉛, ... ✗

4: 13, ⑰, 21, 25, ㉙, ... ✗

6: 13, ⑲, 25, ㉛, ... ✗

8: 13, 21, ㉙, 37 ✓

12: 13, 25, 37 ✗

24: 13, 37 ✗

Only the common difference 8 gives a sequence with exactly one prime between 13 and 37. That prime is **29**.

82. The common difference of a sequence of integers that includes 27 and 34 must be a factor of $34-27=7$.

The common difference of a sequence of integers that includes 34 and 49 must be a factor of $49-34=15$.

The only number that is a factor of both 7 and 15 is 1. So, the greatest possible common difference is **1**.

83. The common difference of a sequence of integers that includes 20 and 32 must be a factor of $32-20=12$.

The common difference of a sequence of integers that includes 32 and 50 must be a factor of $50-32=18$.

So, the common difference is a factor of both 12 and 18. To make the 10th term as large as possible, we use the greatest common factor of 12 and 18, which is 6.

We also make the 10th term as large as possible by assuming that the smallest given number (20) is the 1st term. The 10th term of an arithmetic sequence with 1st term 20 and common difference 6 is $20+9(6)=20+54=\textbf{74}$.

The entire sequence is shown below:

20, 26, 32, 38, 44, 50, 56, 62, 68, 74

84. The common difference of a sequence of integers that includes 23 and 58 must be a factor of $58-23=35$. Since the terms are non-consecutive, the common difference cannot be 1. The remaining factors of 35 are 5, 7, and 35.

Every term in the sequence can be written as 23 plus or minus some multiple of the common difference.

If the common difference is 5, the smallest positive integer we can have is $23-4(5)=23-20=3$.

If the common difference is 7, the smallest positive integer we can have is $23-3(7)=23-21=2$.

If the common difference is 35, the smallest positive integer we can have is 23.

So, the smallest possible positive integer is **2**.

85. Since the sequence includes the terms 33, 53, and 73, and all terms are positive integers, its common difference must be a factor of $53-33=20$ and a factor of $73-53$, which also is 20.

The factors of 20 are 1, 2, 4, 5, 10, and 20. However, the sequence has seven terms, and we cannot go from 33 to 73 with common difference 1, 2, 4, or 5 using just seven terms. For example, if the common difference is 5, then the sequence must include at least nine terms:

33, 38, 43, 48, 53, 58, 63, 68, 73.

Therefore, the common difference can only be 10 or 20.

If the common difference is 10, then we can have any of the following sequences.

13, 23, 33, 43, 53, 63, 73.

23, 33, 43, 53, 63, 73, 83.

33, 43, 53, 63, 73, 83, 93.

If the common difference is 20, then we can have either of the following sequences.

13, 33, 53, 73, 93, 113, 133.

33, 53, 73, 93, 113, 133, 153.

We circle the terms that appear in at least one of the sequences shown above.

3   ⑬   28   38   ⑥③   ⑮③

86. We begin by writing the eight terms, using blanks for the missing units digits.

3_, 3_, 3_, 4_, 4_, 5_, 5_, 5_

If the common difference is 5 or greater, then the third term is at least 10 more than the first term. Since the first and third terms are both in the 30's, they must be less than 10 apart. So, the common difference must be less than 5.

The 3rd term in the sequence is at most 39. If the common difference is 3 or less, then the 6th term is at most $39+3(3)=48$. However, the 6th term has a tens digit of 5. So, the common difference cannot be 3 or less.

Therefore, the common difference must be **4**.

There are two sequences we can make with a common difference of 4:

30, 34, 38, 42, 46, 50, 54, 58,

31, 35, 39, 43, 47, 51, 55, 59.

87. The units digits of the middle two terms are 4 and 9. So, the common difference of this sequence must have units digit 5. We use this fact to fill in the missing units digits.

_4, 3 **9**, _4, _9, 8 **4**, _9

Then, only a common difference of 15 takes us from 39 to 84 in three steps.

We use the common difference of 15 to fill in the remaining blanks as shown.

2 4, 3 9, 5 4, 6 9, 8 4, 9 9

88. The top row contains a 1 and a 4. The only arithmetic sequence of positive integers that can be made with 1 and 4 is (1, 4, 7). So, the top-right square is **7**.

1	4	**7**
5		8
	6	9

The middle row is (2, 5, 8) or (5, 8, 11). Only placing **2** in the middle square also creates an arithmetic sequence in the middle column: (2, 4, 6).

1	4	**7**
5	**2**	8
	6	9

The bottom row is (3, 6, 9) or (6, 9, 12). Only placing **3** in the bottom-left square also creates an arithmetic sequence in the left column: (1, 3, 5).

1	4	**7**
5	**2**	8
3	6	9

89. The top row contains a 2 and a 5. The only arithmetic sequence of positive integers that can be made with 2 and 5 is (2, 5, 8). So, the top-middle square is **8**.

2	**8**	5
7	9	
		8

The left column contains a 2 and a 7. The only arithmetic sequence of positive integers that can be made with 2 and 7 is (2, 7, 12). So, the bottom-left square is **12**.

2	**8**	5
7	9	
12		8

The right column is (2, 5, 8) or (5, 8, 11). Only an **11** in the middle-right square also creates an arithmetic sequence in the middle row: (7, 9, 11).

2	**8**	5
7	9	**11**
12		8

The middle column is (7, 8, 9) or (8, 9, 10). Only a **10** in the bottom-middle square also creates an arithmetic sequence in the bottom row: (8, 10, 12).

2	**8**	5
7	9	**11**
12	**10**	8

90. Step 1:   Step 2:

	25	
3	**29**	16
1		14

	25	
3	**29**	16
1	**27**	14

Step 3: The left column is (1, 2, 3) or (1, 3, 5). If the top-left square is 2, then the only arithmetic sequence we can make in the top row is (2, 25, 48), so the top-right square would be 48. However, (14, 16, 48) is not an arithmetic sequence.

So, the top-left square is **5**.

2	25	48
3	29	16
1	27	14

5	25	
3	**29**	16
1	**27**	14

Step 4: The top row is (5, 15, 25) or (5, 25, 45). Only a **15** in the top-right square also gives an arithmetic sequence in the right column: (14, 15, 16).

5	25	**15**
3	**29**	16
1	**27**	14

91. Step 1:

	60	
1	32	**63**
99		77

Step 2:

	60	
1	32	**63**
99	**88**	77

Step 3:

197	60	334
1	32	**63**
99	88	77

Step 4:

50	60	**70**
1	32	**63**
99	**88**	77

50	60	
1	32	**63**
99	**88**	77

92. Step 1:

17		22
51	36	
34		23

Step 2:

17		22
51	36	**21**
34		23

Step 3:

17	12	22
51	36	**21**
34	28	23

24 or 50

Step 4:

17	**27**	22
51	36	**21**
34	**45**	23

17	**27**	22
51	36	**21**
34		23

93. Step 1:

6	**24**	15
11	33	
		8

Step 2:

6	**24**	15
11	33	**22**
		8

Step 3:
The left column is
(1, 6, 11) or (6, 11, 16).
The middle column is
(15, 24, 33) or (24, 33, 42).

6	**24**	15
11	33	**22**
		8

1 or 16   15 or 42

Only a **1** in the bottom-left square and a **15** in the bottom-middle square gives an arithmetic sequence in the bottom row: (1, 8, 15).

6	**24**	15
11	33	**22**
1	**15**	8

94. Step 1:

	15	
	2	
5	**28**	
79		20

Step 2:

	15	
	2	
5	**28**	51
79		20

Step 3:

	15	
	2	**82**
5	**28**	51
79		20

Step 4:

	15	
42	2	**82**
5	**28**	51
79		20

95. Step 1:

7		14
22		
37		18
	84	

Step 2:

7		14
22		
37	**56**	18
	84	

Step 3:
The middle column is
([28], 56, 84) or (56, [70], 84) or (56, 84, [112]).
The right column is
([10], 14, 18) or (14, [16], 18) or (14, 18, [22]).

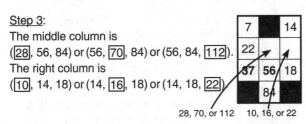

28, 70, or 112   10, 16, or 22

Only a **28** in the middle column and a **16**
in the right column make an arithmetic
sequence in the row with 22: (16, 22, 28).

96. The middle column is (5, [12], 19) or
(5, 19, [33]). Only a **33** also makes an
arithmetic sequence in the row with
11 and 55: (11, 33, 55).

The left column is
([15], 35, 55) or (35, [45], 55) or (35, 55, [75]).
The right column is
([7], 9, 11) or (9, [10], 11) or (9, 11, [13]).

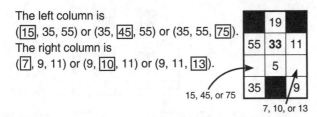

15, 45, or 75

7, 10, or 13

Only a **15** in the left column and a **10** in the
right column makes an arithmetic sequence
in the row with 5: (5, 10, 15).

97. In the right column, the common difference is 2 or 1.
So, the smallest term that could be in the right column is
$32 - 2(2) = 28$, and the largest term that could be in the
right column is $34 + 2(2) = 38$.

The difference between the given terms in the bottom
row is $91 - 1 = 90$. So, we have the following possibilities
for the common difference:

- Common difference 90 gives (1, 91, [181], [271]).
- Common difference $90 \div 2 = 45$ gives (1, [46], 91, [136]).
- Common difference $90 \div 3 = 30$ gives (1, [31], [61], 91).

The rightmost square of the bottom row is between 28
and 38. Only (1, [31], [61], 91) includes a term between 28
and 38, so the rightmost square in the bottom row is **31**,
and the other empty square is **61**. Then, the remaining
empty square in the right column is **33**.

In the left column, the difference
between 8 and 1 is 7. So, the common
difference is 7. Therefore, the numbers
in the left column are (1, 8, [15], [22]),
and the top-left square is 15 or 22.

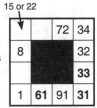

15 or 22

In the top row, there is no arithmetic sequence we can
make with 22, 34, 72, and one other term. So, the top
top-left square is **15**, and the square below the 8 is **22**.

Then, we can only place a **53** in the top row to make an
arithmetic sequence: (15, 34, 53, 72).

98. The difference between the given terms in the left column
is $19 - 9 = 10$. So, the common difference in the left
column is either 10 or $10 \div 2 = 5$. Therefore, every number
in the left column has units digit 9 or 4.

In the bottom row, we have $23 - 7 = 16$. So, the common
difference is either 16 or $16 \div 2 = 8$. This gives the
following possibilities for the numbers in the bottom row:

(7, 23, [39], [55]) or (7, [15], 23, [31]).

Of these numbers, only 39 has a units digit of 9 or 4, so
39 goes in the bottom-left square, and **55** goes in the
remaining empty square in the bottom row.

Then, the missing term in the left column is **29**.

In the top row, the difference between 46 and 29 is 17.
So, the common difference is 17. This gives the following
possibilities for the numbers in the top row:

([12], 29, 46, [63]) or (29, 46, [63], [80]).

So, the number in the top-right square is 12, 63, or 80.
Of these, only 12 can be used to make an arithmetic
sequence in the right column with 23 and 45. So, the
top-right square is **12**, and the sequence in the right
column is (12, 23, **34**, 45).

We place a **63** in the top row to complete the puzzle.

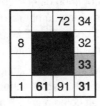

99. We use the strategies discussed in previous problems to complete the puzzle as shown.

2	4	5	3
16	■	■	13
30	■	■	8
44	57	31	18

100. In the middle column, the difference between 22 and 17 is 5. So, the common difference is 5. Therefore, the missing numbers in the middle column are 7 and 12.

In the middle row, we can go from 1 to 4 with a common difference of 1 or 3. However, we cannot go from 1 to 13 in just five terms with a common difference of 1. So, the common difference is 3. Therefore, the missing numbers in the middle row are 7 and 10.

Since 7 is the only number that can go in both the middle row and the middle column, we place a **7** in the center square. Then, we place the **12** and **10** as shown.

Then, we use the strategies discussed in previous problems to complete the puzzle as shown.

101. In the left column, the common difference is 22−15 = 7. So, the missing numbers in this column are 8 and 29.

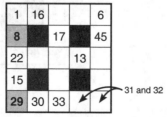

8 and 29

Since 8 cannot be part of the 5-term arithmetic sequence in the bottom row that includes 30 and 33, the bottom-left square is **29**, and the remaining empty square in the left column is **8**.

Then, the common difference in the bottom row is 1, and the missing numbers in the bottom row are 31 and 32.

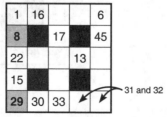

31 and 32

Since 31 cannot be part of the 5-term arithmetic sequence in the right column that includes 6 and 45, the bottom-right square is **32**, and the remaining empty square in the bottom row is **31**.

Then, the common difference in the right column is 45−32 = 13, and the missing numbers are 19 and 58.

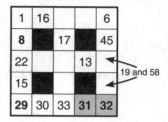

19 and 58

Since 58 cannot be part of the 5-term arithmetic sequence in the middle row that includes 13 and 22, the rightmost square in the middle row is **19**, and the remaining empty square in the right column is **58**.

1	16			6
8	■	17	■	45
22			13	19
15	■		■	58
29	30	33	31	32

We use the previous strategies to complete the puzzle as shown.

10 or 25

11 and 21

1	16	21	11	6
8	■	17	■	45
22	16	25	13	19
15	■	29	■	58
29	30	33	31	32

102. In the bottom row, the common difference is 35−28 = 7. Since 28 and 35 are multiples of 7, every number in the bottom row is a multiple of 7.

In the left column, the common difference is a factor of 43−16 = 27. We can eliminate 1, 3, and 27. So, the common difference in the left column is 9. Therefore, the missing numbers in the left column are 25, 31, and either 7 or 52. Only 7 is a multiple of 7, so the square in the left column and bottom row is **7**.

Then, the remaining empty squares in the bottom row are 14 and 21, and the remaining empty squares in the left column are 25 and 34.

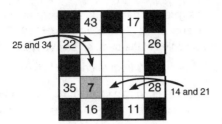

We consider the empty square above the 11. We cannot make a 5-term arithmetic sequence with the numbers 11, 17, and 21. So, we place **14** above the 11, and place **21** to the left of the 14.

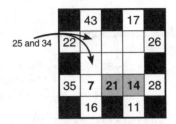

If we place 34 to the right of the 22, then

- the common difference in the top row is 4, and
- one of the remaining empty squares in the top row is 30, and the other is 18 or 38.

However, we cannot place 18, 30, or 38 in the right column. So, 34 cannot go to the right of 22. So, we place **25** to the right of the 22, and place **34** beneath the 25.

Therefore, the common difference in the top row is 1, and the missing numbers in the top row are 23 and 24.

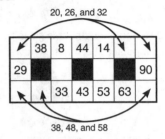

We use the previous strategies to complete puzzle as shown.

103. In the top row, the common difference is a factor of 14−8 = 6. We cannot go from 8 to 44 in seven terms using a common difference of 1, 2, or 3. So, the common difference is 6.

Therefore, the missing numbers in the top row are 20, 26, and 32.

In the bottom row, the common difference is a factor of 43−33 = 10. We cannot go from 33 to 63 in seven terms using a common difference of 1 or 2. So, the common

---

difference in the bottom row is 5 or 10. If the common difference is 10, then every number in the bottom row has units digit 3. However, we cannot make an arithmetic sequence in the right column with 90, a number with units digit 3, and one of (20, 26, or 32). So, the common difference in the bottom row is 5.

Therefore, the missing numbers in the bottom row are 38, 48, and 58.

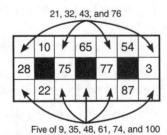

We fill the blanks shown below in the only way that gives an arithmetic sequence in the left and right columns.

20	38	8	44	14		26
29						90
38		33	43	53	63	58

Then, we complete the puzzle as shown.

20	38	8	44	14	32	26
29		58		92		90
38	48	33	43	53	63	58

104. In the top row, the common difference is 65−54 = 11. So, the missing numbers in the top row are 21, 32, 43, and 76.

In the bottom row, the common difference is a factor of 87−22 = 65. So, the common difference is 1, 5, 13, or 65. We cannot go from 22 to 87 in seven terms using a common difference of 1 or 5, and we can eliminate 65. So, the common difference in the bottom row is 13.

Therefore, the missing numbers in the bottom row are *five* of the following: 9, 35, 48, 61, 74, and 100.

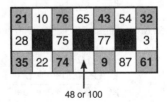

We fill the blanks shown below in the only way that creates an arithmetic sequence in each column.

21	10	76	65	43	54	32
28		75		77		3
35	22	74		9	87	61

48 or 100

The remaining empty square can only be 48, so we complete the puzzle as shown.

21	10	76	65	43	54	32
28	■	75	■	77	■	3
35	22	74	48	9	87	61

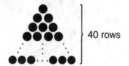

SEQUENCES
Triangular Numbers
22-23

105. The 40th triangular number is given by the number of dots in the triangular pattern with 40 rows.

$$\left.\begin{matrix} \bullet \\ \bullet\bullet \\ \bullet\bullet\bullet \\ \vdots \\ \bullet\bullet\cdots\bullet\bullet \end{matrix}\right\} 40 \text{ rows}$$

This is equal to the sum $1+2+3+\cdots+38+39+40$.

To compute this sum, we add two copies of it, with one copy written in reverse.

$$\begin{array}{r} 1 + 2 + 3 + \cdots + 38 + 39 + 40 \\ 40 + 39 + 38 + \cdots + 3 + 2 + 1 \\ \hline 41 + 41 + 41 + \cdots + 41 + 41 + 41 \end{array}$$

This gives 40 pairs of numbers that each sum to 41. So, the sum of all 40 pairs is $40 \cdot 41 = 1,640$. However, this is the total of *two* copies of the sum we wish to compute. So, the value of *one* copy is $1,640 \div 2 = 820$.

So, the 40th triangular number is **820**.

106. To simplify $1+2+3+\cdots+(n-2)+(n-1)+n$, we use our strategy from the previous problem. We add two copies of this sum, with one copy written in reverse.

$$\begin{array}{r} 1 + 2 + 3 + \cdots + (n-2) + (n-1) + n \\ n + (n-1) + (n-2) + \cdots + 3 + 2 + 1 \\ \hline (n+1)+(n+1)+(n+1)+\cdots+(n+1)+(n+1)+(n+1) \end{array}$$

This gives n pairs of numbers that each sum to $n+1$. So, the sum of all n pairs is $n(n+1)$. Since this is the total of two copies of the sum, the value of one copy is $\frac{n(n+1)}{2}$.

So, a simpler expression for the nth triangular number is $\frac{n(n+1)}{2}$. You may have also distributed the n in the numerator to get $\frac{n^2+n}{2}$.

107. Factoring a 2 from each term in the sum, we have

$$2+4+6+8+10+12+14+16+18+20+22+24$$
$$=2(1+2+3+4+5+6+7+8+9+10+11+12).$$

This is 2 times the sum of the first twelve positive integers, or 2 times the 12th triangular number.

The 12th triangular number is $\frac{12 \cdot 13}{2}$. So, 2 times the 12th triangular number is $2 \cdot \frac{12 \cdot 13}{2} = 12 \cdot 13 = \mathbf{156}$.

— *or* —

We add two copies of the sum, with one copy written in reverse.

$$\begin{array}{r} 2 + 4 + 6 + \cdots + 20 + 22 + 24 \\ 24 + 22 + 20 + \cdots + 6 + 4 + 2 \\ \hline 26 + 26 + 26 + \cdots + 26 + 26 + 26 \end{array}$$

This gives 12 pairs of numbers that each sum to 26. So, the sum of all 12 pairs is $12 \cdot 26 = 312$. This is the total of two copies of the sum, so one copy is $312 \div 2 = \mathbf{156}$.

108. Factoring an 8 from each term in the sum, we have

$$88+80+72+64+56+48+40+32+24+16+8$$
$$=8(11+10+9+8+7+6+5+4+3+2+1).$$

This is 8 times the sum of the first eleven positive integers, or 8 times the 11th triangular number.

The 11th triangular number is $\frac{11 \cdot 12}{2}$. So, 8 times the 11th triangular number is $8 \cdot \frac{11 \cdot 12}{2} = 4 \cdot 11 \cdot 12 = \mathbf{528}$.

— *or* —

We add two copies of the sum, with one copy written in reverse.

$$\begin{array}{r} 88 + 80 + 72 + \cdots + 24 + 16 + 8 \\ 8 + 16 + 24 + \cdots + 72 + 80 + 88 \\ \hline 96 + 96 + 96 + \cdots + 96 + 96 + 96 \end{array}$$

This gives 11 pairs of numbers that each sum to 96. So, the sum of all 11 pairs is $11 \cdot 96 = 1,056$. This is the total of two copies of the sum, so one copy is $1,056 \div 2 = \mathbf{528}$.

109. Factoring a 6 from each term in the sum, we have

$$6+12+18+\cdots+588+594+600$$
$$=6(1+2+3+\cdots+98+99+100).$$

This is 6 times the 100th triangular number. So, the sum is equal to $6 \cdot \frac{100 \cdot 101}{2} = 3 \cdot 100 \cdot 101 = \mathbf{30,300}$.

110. Writing each mixed number as a fraction, then factoring $\frac{4}{7}$ from each term in the sum, we have

$$\frac{4}{7}+1\frac{1}{7}+1\frac{5}{7}+2\frac{2}{7}+\cdots+11\frac{3}{7}$$
$$=\frac{4}{7}+\frac{8}{7}+\frac{12}{7}+\frac{16}{7}+\cdots+\frac{80}{7}$$
$$=\frac{4}{7}(1+2+3+4+\cdots+20).$$

This is $\frac{4}{7}$ times the 20th triangular number. So, the sum is equal to

$$\frac{4}{7} \cdot \frac{20 \cdot 21}{2} = \frac{4}{\cancel{7}_1} \cdot \frac{\cancel{20}^{10} \cdot \cancel{21}^{3}}{\cancel{2}_1} = \mathbf{120}.$$

111. We add two copies of this sum, with one copy written in reverse.

$$\begin{array}{r} 51 + 52 + 53 + \cdots + 98 + 99 + 100 \\ 100 + 99 + 98 + \cdots + 53 + 52 + 51 \\ \hline 151 + 151 + 151 + \cdots + 151 + 151 + 151 \end{array}$$

There are 50 terms in $51+52+53+\cdots+98+99+100$. So, we have 50 pairs of numbers that each sum to 151. Therefore, the sum of all 50 pairs is $50 \cdot 151 = 7,550$. This is the total of two copies of the sum, so one copy is $7,550 \div 2 = \mathbf{3,775}$.

— *or* —

We notice that $51+52+53+\cdots+98+99+100$ is the sum of the first 100 positive integers *minus* the sum of the first 50 positive integers.

$$\begin{array}{r} (1+2+3+\cdots+48+49+50+51+52+53+\cdots+98+99+100) \\ -(1+2+3+\cdots+48+49+50) \\ \hline 51+52+53+\cdots+98+99+100 \end{array}$$

This is equal to the 100th triangular number minus the 50th triangular number, which we compute as follows:

$$\frac{100 \cdot 101}{2} - \frac{50 \cdot 51}{2} = \frac{10,100}{2} - \frac{2,550}{2}$$
$$= 5,050 - 1,275$$
$$= \mathbf{3,775}.$$

112. The 25th perfect square is 25^2.
The 25th triangular number is $\frac{25 \cdot 26}{2}$.

So, the 25th term in Ariana's sequence is

$$25^2 - \frac{25 \cdot 26}{2} = 25^2 - 25 \cdot 13$$
$$= 25 \cdot 25 - 25 \cdot 13$$
$$= 25(25 - 13)$$
$$= 25(12)$$
$$= \mathbf{300}.$$

— *or* —

The following diagram shows the first several terms in Ariana's sequence. In each figure, we subtract the nth triangular number (given by the dots in gray) from the nth perfect square (given by the total dots). The number of remaining black dots gives the nth term in Ariana's sequence.

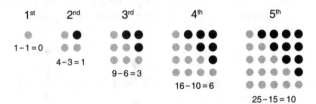

After the 1st term, the numbers in Ariana's sequence are the triangular numbers! For example, her 2nd term (1) is the 1st triangular number. Her 3rd term (3) is the 2nd triangular number, and so on.

Continuing this pattern, Ariana's 25th term is equal to the 24th triangular number, which is $\frac{24 \cdot 25}{2} = 12 \cdot 25 = \mathbf{300}$.

113. The 997th triangular number is the sum of the first 997 positive integers. The 1,002nd triangular number is the sum of all of the same integers, plus the integers from 998 to 1,002. So, the difference between the 1002nd and 997th triangular numbers is the sum of the integers from 998 to 1002.

$$1+2+3+\cdots+995+996+997+998+\cdots+1,002$$
$$-(1+2+3+\cdots+995+996+997)$$
$$\overline{\qquad\qquad\qquad\qquad 998+\cdots+1,002}$$

The terms in $998+999+1,000+1,001+1,002$ balance around 1,000. So, their average is 1,000, and their sum is therefore $5 \cdot 1,000 = \mathbf{5,000}$.

114. The kth triangular number is equal to $\frac{k(k+1)}{2}$. We are told that the kth triangular number is 465. So, we write an equation:

$$\frac{k(k+1)}{2} = 465.$$

We can multiply both sides of the equation by 2 to get rid of the denominator on the left hand side. This gives

$$k(k+1) = 930.$$

Since k is the position number of a term in the sequence, we know that k is a positive integer. So, we look for two consecutive positive integers, k and $k+1$, whose product is 930. We know that $900 = 30^2$, so 30 is a reasonable guess for k.

If $k = 30$, then $k(k+1) = 30(31) = 930$. ✓
So, $k = \mathbf{30}$. The 30th triangular number is 465.

115. We add two copies of this series with one copy written in reverse.

$$4 + 5 + 6 + \cdots + 14 + 15 + 16$$
$$16 + 15 + 14 + \cdots + 6 + 5 + 4$$
$$\overline{20 + 20 + 20 + \cdots + 20 + 20 + 20}$$

This gives 13 pairs of numbers that each sum to 20. The sum of all 13 pairs is $13 \cdot 20 = 260$. Since this is the sum of two copies of the series, one copy equals $260 \div 2 = \mathbf{130}$.

— *or* —

We find the average of the terms in the series and multiply by the number of terms.

Since the terms in the series form an arithmetic sequence, they balance around the median. Therefore, their average is equal to the median.

Since the median is 10, the average is also 10. The sum of 13 terms with an average of 10 is $13 \cdot 10 = \mathbf{130}$.

116. We find the average of the terms in the series and multiply by the number of terms.

Since the terms in the series form an arithmetic sequence, they balance around the median. Therefore, their average is equal to the median.

Since the median is 55, the average is also 55. The sum of 8 terms with an average of 55 is $8 \cdot 55 = \mathbf{440}$.

117. We find the average of the terms in the series and multiply by the number of terms.

Since the terms in the series form an arithmetic sequence, they balance around the median. Therefore, their average is equal to the median, which is 54.

The sum of 9 terms with an average of 54 is $9 \cdot 54 = \mathbf{486}$.

118. The 15 terms in this series balance around their median, which is 1. Therefore, the average of the terms is 1.

The sum of 15 terms with an average of 1 is $15 \cdot 1 = \mathbf{15}$.

— *or* —

Each negative term in the sum cancels its corresponding positive term. For example, -6 + 6 = 0, -5 + 5 = 0, and so on.

After canceling the pairs of terms that sum to zero, we are left with $0+7+8 = \mathbf{15}$.

119. The 10 terms in this series balance around their median, which is $\frac{35+41}{2} = \frac{76}{2} = 38$. Therefore, the average of the terms is 38.

The sum of 10 terms with an average of 38 is $10\cdot38 = \mathbf{380}$.

120. Converting the terms to a sequence of consecutive integers makes them easier to count.

The common difference in the sequence is 4.

$$9,\ 13,\ 17,\ 21,\ 25,\ ...,\ 45,\ 49,\ 53,\ 57,\ 61.$$

Subtracting 5 from each term gives the multiples of 4.

$$4,\ 8,\ 12,\ 16,\ 20,\ ...,\ 40,\ 44,\ 48,\ 52,\ 56.$$

Then, dividing each term by 4, we have

$$1,\ 2,\ 3,\ 4,\ 5,\ ...,\ 10,\ 11,\ 12,\ 13,\ 14.$$

These steps did not change the number of terms in the sequence. So, there are **14** terms.

121. Since there are 14 terms in the sequence, the median is the number halfway between the 7th and 8th terms.

The 7th term is $9+6(4) = 33$ and the 8th term is $9+7(4) = 37$. So, the median is $\frac{33+37}{2} = \frac{70}{2} = \mathbf{35}$.

122. Since the terms form an arithmetic sequence, their average is equal to their median, 35. The sum of 14 terms with an average of 35 is $14\cdot35 = \mathbf{490}$.

123. In an arithmetic sequence, the average is equal to the median. In this sequence, the median is the number halfway between the two middle terms. Since the terms in an arithmetic sequence balance around the median, the number halfway between the two middle terms is the same as the number halfway between the first term and the last term.

$$9,\ 13,\ 17,\ 21,\ 25,\ 29,\ 33,\ 37,\ 41,\ 45,\ 49,\ 53,\ 57,\ 61$$

The number halfway between 9 and 61 is $\frac{9+61}{2} = \frac{70}{2} = 35$. **So, we can compute the average of this arithmetic sequence, or of any arithmetic sequence, by taking the average of the first term and the last term.**

124. Since the middle term of the sequence is 43 and it is an arithmetic sequence, the average of the terms in the sequence is also 43.

In Problem 123, we learned that the average of an arithmetic sequence is equal to the average of the first and last terms. So, the average of the first and last terms is 43. Therefore, the sum of the first and last terms is $43\cdot2 = \mathbf{86}$.

125. The average of the terms in an arithmetic sequence with first term 19 and last term 81 is $\frac{19+81}{2} = \frac{100}{2} = 50$.

The sum of 30 terms with an average of 50 is $30\cdot50 = \mathbf{1,500}$.

126. The first term of the sequence is 24 and the common difference is 4. So, the 100th term in the sequence is $24+99(4) = 24+396 = 420$.

The average of the terms is the same as the average of the first and last terms: $\frac{24+420}{2} = \frac{444}{2} = 222$. So, the sum of all 100 terms in the sequence is $100\cdot222 = \mathbf{22,200}$.

127. To compute the sum of the series, we multiply the average of the terms by the number of terms.

Since the terms in $15+21+27+\cdots+207$ form an arithmetic sequence, the average of the terms is equal to the average of the first and last terms: $\frac{15+207}{2} = \frac{222}{2} = 111$.

To find the number of terms, we convert the terms to a sequence of consecutive integers. The common difference of the terms is 6.

$$15,\ 21,\ 27,\ ...,\ 207.$$

Subtracting 9 from each term, we get the multiples of 6.

$$6,\ 12,\ 18,\ ...,\ 198.$$

Dividing each term by 6 gives

$$1,\ 2,\ 3,\ ...,\ 33.$$

So, there are 33 terms in the series. The sum of 33 terms with an average of 111 is $33\cdot111 = \mathbf{3,663}$.

Gumball Patterns   26-27

128. The first figure in the pattern has 5 gumballs. Then, each figure has 4 more gumballs than the figure that came before it.

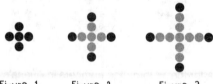

Figure 1   Figure 2   Figure 3

The number of gumballs in each figure of the pattern makes an arithmetic sequence with first term 5 and common difference 4.

So, to make the 8th figure in the pattern, $5+7(4) = 5+28 = \mathbf{33}$ gumballs are needed.

— *or* —

Each figure has a gumball in the center, plus four "arms." The number of gumballs in each arm is equal to the figure number. For example, the 1st figure has 1 gumball per arm, the 2nd figure has 2 gumballs per arm, the 3rd figure has 3 gumballs per arm, and so on.

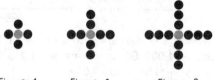

Figure 1   Figure 2   Figure 3

So, the 8th figure has four arms with 8 gumballs each, plus one gumball in the center. This gives a total of $4(8)+1 = \mathbf{33}$ gumballs.

129. To make the nth figure in the pattern, we begin with the 5 gumballs in the first figure and add 4 gumballs ($n-1$) times.

So, the number of gumballs needed to make the nth figure is $5+(n-1)4$. Distributing the 4 and combining like terms, this simplifies to $5+4n-4=\boldsymbol{4n+1}$.

— or —

The nth figure in the pattern has 4 arms with n gumballs each, plus one gumball in the center. So, the nth figure has $\boldsymbol{4n+1}$ gumballs.

130. From the previous problem, we know that the kth figure in the pattern uses $4k+1$ gumballs. We are told that 221 gumballs are used in the kth figure, so

$$4k+1=221.$$

Subtracting 1 from both sides of the equation gives

$$4k=220.$$

Dividing both sides of the equation by 4, we have $k=\boldsymbol{55}$. The 55th figure in the pattern uses 221 gumballs.

131. The number of gumballs in each figure is given by a sum of consecutive odd numbers. For example, the number of gumballs in the 4th figure is the sum of the first four odd numbers: $1+3+5+7=16$.

Figure 4

So, the number of gumballs in the 30th figure is the sum of the first 30 odd numbers. In Problem 30, we learned that the nth odd number is $2n-1$. So, the 30th odd number is $2(30)-1=59$.

Therefore, the number of gumballs in the 30th figure is equal to $1+3+5+\cdots+55+57+59$. This is an arithmetic series! The 30 terms in this series have an average of $\frac{1+59}{2}=30$, so the sum of all of the terms is $30\cdot30=\boldsymbol{900}$.

— or —

We can rearrange the gumballs in each figure to make a square, as shown in the diagram below.

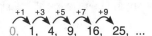

Figure 1   Figure 2   Figure 3   Figure 4

The 1st figure has $1^2=1$ gumball.
The 2nd figure has $2^2=4$ gumballs.
The 3rd figure has $3^2=9$ gumballs.
The 4th figure has $4^2=16$ gumballs.

Continuing this pattern, the 30th figure has $30^2=\boldsymbol{900}$ gumballs.

You may recall from the Perfect Squares chapter of Beast Academy 3B that the sum of the first n odd numbers is equal to n^2. This is why the number of gumballs in each figure is always a perfect square.

$$\overset{+1}{\frown}\ \overset{+3}{\frown}\ \overset{+5}{\frown}\ \overset{+7}{\frown}\ \overset{+9}{\frown}$$
$$0,\ 1,\ 4,\ 9,\ 16,\ 25,\ \ldots$$

132. To get from the 1st figure to the 2nd, we add 4 gumballs. To get to each figure after that, we add 2 more gumballs than we added the previous time.

Figure 1   Figure 2   Figure 3   Figure 4

So, the number of gumballs in the 10th figure is

$$1+(4+6+8+10+12+14+16+18+20).$$

The sum in parentheses is an arithmetic series with 9 terms that have an average of $\frac{20+4}{2}=12$. So, the sum in parentheses equals $9\cdot12=108$.

Therefore, the number of gumballs in the 10th figure is $1+(4+6+8+10+12+14+16+18+20)=1+108=\boldsymbol{109}$.

— or —

If we ignore the gumballs on the right half of each figure, the remaining gumballs (shown in black below) make a pattern of triangular numbers.

Figure 1   Figure 2   Figure 3   Figure 4

Similarly, if we ignore the gumballs on the left half of each figure, the remaining gumballs form the same pattern of triangular numbers.

Figure 1   Figure 2   Figure 3   Figure 4

So, we can count the total number of gumballs in each figure by adding the gumballs on the left plus the gumballs on the right. However, this counts the middle gumball twice, so we must subtract 1 at the end.

So, the number of gumballs in the 10th figure of the pattern is 2 times the 10th triangular number, minus 1. This gives $2\cdot\frac{10\cdot11}{2}-1=10\cdot11-1=110-1=\boldsymbol{109}$.

133. We begin by finding an expression for the number of gumballs in the kth figure of the pattern.

Each figure in the pattern forms a rectangular grid of gumballs. We can summarize the dimensions of each rectangular grid in the following table.

Figure #	1	2	3	4
Dimensions	1 by 2	2 by 3	3 by 4	4 by 5

Following this pattern, the kth figure in the sequence forms a k-by-$(k+1)$ grid, giving $k(k+1)$ total gumballs.

We want to know when $k(k+1)=420$. Since k is a positive integer, we look for two consecutive positive integers, k and $k+1$, whose product is 420.

420 is close to $400=20^2$, so $k=20$ is a reasonable guess. If $k=20$, then $k(k+1)=20(21)=420$. ✓

So, $k=\boldsymbol{20}$. The 20th figure uses 420 gumballs.

134. The number of black gumballs and the number of gray gumballs in each figure are always perfect squares. It is easier to see why this is true if we rotate each figure 45 degrees, as shown below.

Figure 1   Figure 2   Figure 3   Figure 4

- Figure 1 has a "1-by-1 square" of black gumballs, for a total of $1^2 = 1$ gumball.

- Figure 2 has a 2-by-2 square of gray gumballs and a "1-by-1 square" of black gumballs, for a total of $2^2 + 1^2 = 5$ gumballs.

- Figure 3 has a 3-by-3 square of black gumballs and a 2-by-2 square of gray gumballs, for a total of $3^2 + 2^2 = 13$ gumballs.

- Figure 4 has a 4-by-4 square of gray gumballs and a 3-by-3 square of black gumballs, for a total of $4^2 + 3^2 = 25$ gumballs.

Continuing this pattern, the nth figure has an n-by-n square of gray gumballs, and an $(n-1)$-by-$(n-1)$ square of black gumballs, for a total of $\boldsymbol{n^2 + (n-1)^2}$ gumballs.

Note: There are several equivalent ways to write this expression. If you solved this problem a different way, you may have gotten $\boldsymbol{2n^2 - 2n + 1}$ or $\boldsymbol{2(n^2 - n) + 1}$.

SEQUENCES
The Fibonacci Sequence   28-29

135. The first five terms of the Fibonacci sequence are given.

$$1, 1, 2, 3, 5, \_\_, \_\_, \_\_, \_\_, \_\_$$

To find the 6th term, we add the 4th and 5th terms: $3 + 5 = 8$.

$$1, 1, 2, 3, 5, \mathbf{8}, \_\_, \_\_, \_\_, \_\_$$

The 7th term is the sum of the 5th and 6th terms: $5 + 8 = 13$.
The 8th term is the sum of the 6th and 7th terms: $8 + 13 = 21$.
The 9th term is the sum of the 7th and 8th terms: $13 + 21 = 34$.
The 10th term is the sum of the 8th and 9th terms: $21 + 34 = 55$.

$$1, 1, 2, 3, 5, \mathbf{8}, \mathbf{13}, \mathbf{21}, \mathbf{34}, \mathbf{55}$$

136. The 20th term of the Fibonacci sequence is the sum of the 18th term and the 19th term. The 19th term is 4,181 and the 20th term is 6,765. If we call the 18th term n, then we have the equation $n + 4,181 = 6,765$.

So, $n = 6,765 - 4,181 = 2,584$. The 18th term is **2,584**.

137. We list the first few terms of the Fibonacci sequence and consider whether they are even or odd.

$$1, \quad 1, \quad 2, \quad 3, \quad 5, \quad 8, \quad 13, \quad 21, \quad 34, \dots$$
$$\text{odd} \quad \text{odd} \quad \text{even} \quad \text{odd} \quad \text{odd} \quad \text{even} \quad \text{odd} \quad \text{odd} \quad \text{even}$$

We see that the numbers follow the repeating pattern of (odd, odd, even). Since

$$\text{odd} + \text{odd} = \text{even},$$
$$\text{odd} + \text{even} = \text{odd, and}$$
$$\text{even} + \text{odd} = \text{odd},$$

this pattern will continue forever.

So, the terms with a position number of 3, 6, 9, 12, 15, and so on, are all even. These are the terms whose position number is a multiple of 3. All of the other terms are odd.

Since 100 is not a multiple of 3, the 100th term is odd.

138. After the first two terms, each term in Grogg's sequence is the sum of the two previous terms. So, if we call the second term of Grogg's sequence a, then the third term in his sequence is $18 + a$.

$$18, a, 18+a, 76, \dots$$

The fourth term, 76, is the sum of the two previous terms. This gives the equation

$$a + (18 + a) = 76.$$

Combining like terms, we have

$$2a + 18 = 76.$$

Subtracting 18 from both sides of the equation gives $2a = 58$. Dividing both sides by 2, we have $a = 29$. So, the second term in Grogg's sequence is **29**.

Check: If the first term is 18 and the second term is 29, then the third term is $18 + 29 = 47$ and the fourth term is $47 + 29 = 76$.

$$18, 29, 47, 76 \checkmark$$

139. To get to the 3rd step, Hoppy must take one of the following paths:

- Three 1-step hops from the base.
  $1 + 1 + 1 = 3$

- A 1-step hop, then a 2-step hop.
  $1 + 2 = 3$

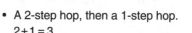

- A 2-step hop, then a 1-step hop.
  $2 + 1 = 3$

There is no other way to reach the 3rd step. So, Hoppy can go from the base of the steps to the 3rd step in **3** ways.

140. Counting the ways that Hoppy can reach the 4th step can get complicated, so we organize our work as shown:

- Four 1-step hops from the base.
  $1 + 1 + 1 + 1 = 4$.

- Two 1-step hops, then a 2-step hop.
  $1 + 1 + 2 = 4$.

- One 1-step hop, then a 2-step hop, then a 1-step hop.
  $1 + 2 + 1 = 4$.

- One 2-step hop, then two 1-step hops.
  $2 + 1 + 1 = 4$.

- Two 2-step hops.
  $2 + 2 = 4$.

There is no other way to reach the 4th step. So, Hoppy can go from the base to the 4th step in **5** ways.

141. Hoppy can only hop directly to the 5th step with a 1-step hop from the 4th step or a 2-step hop from the 3rd step:

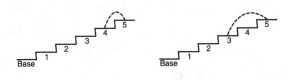

For each of the ways that Hoppy can reach the 4th step, there is exactly one way he can hop straight to the 5th step from there.

For each of the ways that Hoppy can reach the 3rd step, there is exactly one way he can hop straight to the 5th step from there.

Therefore, the number of ways Hoppy can get to the 5th step is the sum of the number of ways he can get to the 4th step and the number of ways he can get to the 3rd step.

As we learned in the previous problems, Hoppy can get to the 4th step in 5 ways and to the 3rd step in 3 ways.

So, he can get to the 5th step in $5+3=\mathbf{8}$ ways.

The following sums represent the 8 paths Hoppy could take to get to the 5th step.

From the 4th step, we have
$(1+1+1+1+1)$,
$(1+1+2+1)$, $(1+2+1+1)$, $(2+1+1+1)$, $(2+2+1)$.

From the 3rd step, we have
$(1+1+1+2)$, $(1+2+2)$, $(2+1+2)$.

142. We use our answers from the previous three problems to fill in the first three blanks.

$$\underset{1^{st}}{\frac{1}{}}, \underset{2^{nd}}{\frac{2}{}}, \underset{3^{rd}}{\frac{3}{}}, \underset{4^{th}}{\frac{5}{}}, \underset{5^{th}}{\frac{8}{}}, \underset{6^{th}}{\frac{}{}}, \underset{7^{th}}{\frac{}{}}, \underset{8^{th}}{\frac{}{}}, \underset{9^{th}}{\frac{}{}}, \underset{10^{th}}{\frac{}{}}, \cdots$$

To find the number of ways Hoppy can reach the 6th step, we use the strategy from the previous problem. Hoppy can only hop directly to the 6th step by making a 1-step hop from the 5th step, or by making a 2-step hop from the 4th step. So, the number of ways Hoppy can reach the 6th step is the sum of the number of ways he can reach the 5th step and the number of ways he can reach the 4th step.

Hoppy can reach the 5th step in 8 ways and the 4th step in 5 ways. So, he can reach the 6th step in $8+5=13$ ways.

$$\underset{1^{st}}{\frac{1}{}}, \underset{2^{nd}}{\frac{2}{}}, \underset{3^{rd}}{\frac{3}{}}, \underset{4^{th}}{\frac{5}{}}, \underset{5^{th}}{\frac{8}{}}, \underset{6^{th}}{\frac{13}{}}, \underset{7^{th}}{\frac{}{}}, \underset{8^{th}}{\frac{}{}}, \underset{9^{th}}{\frac{}{}}, \underset{10^{th}}{\frac{}{}}, \cdots$$

This same pattern continues for each next step. The number of ways Hoppy can reach Step n is the sum of the number of ways he can reach Step $n-1$ and the number of ways he can reach Step $n-2$. This is the same rule that is used by the Fibonacci sequence!

We continue the pattern to fill in the remaining blanks.

$$\underset{1^{st}}{\frac{1}{}}, \underset{2^{nd}}{\frac{2}{}}, \underset{3^{rd}}{\frac{3}{}}, \underset{4^{th}}{\frac{5}{}}, \underset{5^{th}}{\frac{8}{}}, \underset{6^{th}}{\frac{13}{}}, \underset{7^{th}}{\frac{21}{}}, \underset{8^{th}}{\frac{34}{}}, \underset{9^{th}}{\frac{55}{}}, \underset{10^{th}}{\frac{89}{}}, \cdots$$

143. Counting the number of 2-by-10 arrangements of dominos is tough. So, we first consider smaller rectangles.

There is only 1 way to make a 2-by-1 rectangle:

There are 2 ways to make a 2-by-2 rectangle:

There are 3 ways to make a 2-by-3 rectangle:

As the counting gets more difficult, we look for a way to count all of the arrangements using our previous work.

To make a 2-by-4 arrangement, we can add a vertical domino to any of our 2-by-3 rectangles:

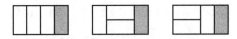

This counts all of the arrangements that have a vertical domino on the right. However, we've missed all of the arrangements that have a pair of horizontal dominos on the right. To get the arrangements that end with a pair of horizontal dominos, we can add two horizontal dominos to the end of either 2-by-2 rectangle:

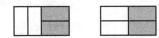

Every 2-by-4 arrangement is either a 2-by-3 arrangement with a vertical domino added at the end, or a 2-by-2 arrangement with two horizontal dominos added at the end. We have counted every arrangement exactly once. So, there are $3+2=5$ ways to make a 2-by-4 rectangle.

Similarly, every 2-by-5 arrangement of dominos is either a 2-by-4 arrangement with a vertical domino added at the end, or a 2-by-3 arrangement with two horizontal dominos added at the end:

So, the number of 2-by-5 rectangles is the sum of the numbers of 2-by-4 rectangles and 2-by-3 rectangles. There are $5+3=8$ ways to make a 2-by-5 rectangle.

The number of 2-by-6 rectangles is the sum of the numbers of 2-by-5 rectangles and 2-by-4 rectangles. There are $8+5=13$ ways to make a 2-by-6 rectangle.

The pattern so far is 1, 2, 3, 5, 8, 13. These are the Fibonacci numbers! Each term is the sum of the two previous terms.

The number of 2-by-7 arrangements is $8+13=21$.
The number of 2-by-8 arrangements is $13+21=34$.
The number of 2-by-9 arrangements is $21+34=55$.
The number of 2-by-10 arrangements is $34+55=89$.

So, there are **89** arrangements of dominos that form a 2-by-10 rectangle.

144. The first two terms of the sequence are 8 and 1, and each next term is half the product of the two previous terms. So, the 3rd term is $\frac{8\cdot1}{2}=\frac{8}{2}=4$.

$$8,\ 1,\ 4,\ \underline{\ \ },\ \underline{\ \ },\ \underline{\ \ },\ \underline{\ \ },\ \underline{\ \ }$$

Then, the 4th term is $\frac{1\cdot4}{2}=\frac{4}{2}=2$.

$$8,\ 1,\ 4,\ 2,\ \underline{\ \ },\ \underline{\ \ },\ \underline{\ \ },\ \underline{\ \ }$$

We continue this pattern to fill in the remaining blanks as shown.

$$8,\ 1,\ 4,\ 2,\ 4,\ 4,\ 8,\ 16$$

145. Each term is the opposite of twice the previous term. So, the term after 5 is $-(2\cdot5)=-10$.

$$5,\ -10,\ \underline{\ \ },\ \underline{\ \ },\ \underline{\ \ },\ \underline{\ \ },\ \underline{\ \ }$$

Then, the term after -10 is $-(2\cdot(-10))=-(-20)=20$.

$$5,\ -10,\ 20,\ \underline{\ \ },\ \underline{\ \ },\ \underline{\ \ },\ \underline{\ \ }$$

We continue this pattern to fill in the remaining blanks as shown.

$$5,\ -10,\ 20,\ -40,\ 80,\ -160,\ 320$$

146. The first two terms are 0 and 1. So, the next term is the reciprocal of $0+1=1$. The reciprocal of 1 is $\frac{1}{1}=1$.

$$0,\ 1,\ 1,\ \underline{\ \ },\ \underline{\ \ },\ \underline{\ \ },\ \underline{\ \ }$$

The next term is the reciprocal of $1+1=2$, which is $\frac{1}{2}$.

$$0,\ 1,\ 1,\ \frac{1}{2},\ \underline{\ \ },\ \underline{\ \ },\ \underline{\ \ }$$

The next term is the reciprocal of $1+\frac{1}{2}=\frac{3}{2}$, which is $\frac{2}{3}$.

$$0,\ 1,\ 1,\ \frac{1}{2},\ \frac{2}{3},\ \underline{\ \ },\ \underline{\ \ }$$

The next term is the reciprocal of $\frac{1}{2}+\frac{2}{3}=\frac{7}{6}$, which is $\frac{6}{7}$.

$$0,\ 1,\ 1,\ \frac{1}{2},\ \frac{2}{3},\ \frac{6}{7},\ \underline{\ \ }$$

The next term is the reciprocal of $\frac{2}{3}+\frac{6}{7}=\frac{32}{21}$, which is $\frac{21}{32}$.

$$0,\ 1,\ 1,\ \frac{1}{2},\ \frac{2}{3},\ \frac{6}{7},\ \frac{21}{32}$$

147. The first term is 65. The sum of the digits of 65 is $6+5=11$, and $11^2=121$. So, the next term is 121.

$$65,\ 121,\ \underline{\ \ },\ \underline{\ \ },\ \underline{\ \ },\ \underline{\ \ },\ \underline{\ \ },\ \underline{\ \ }$$

The sum of the digits of 121 is $1+2+1=4$. So, the next term is $4^2=16$.

$$65,\ 121,\ 16,\ \underline{\ \ },\ \underline{\ \ },\ \underline{\ \ },\ \underline{\ \ },\ \underline{\ \ }$$

The next term is $(1+6)^2=7^2=49$.

$$65,\ 121,\ 16,\ 49,\ \underline{\ \ },\ \underline{\ \ },\ \underline{\ \ },\ \underline{\ \ }$$

The next term is $(4+9)^2=13^2=169$.

$$65,\ 121,\ 16,\ 49,\ 169,\ \underline{\ \ },\ \underline{\ \ },\ \underline{\ \ }$$

The next term is $(1+6+9)^2=16^2=256$.

$$65,\ 121,\ 16,\ 49,\ 169,\ 256,\ \underline{\ \ },\ \underline{\ \ }$$

The next term is $(2+5+6)^2=13^2=169$.

$$65,\ 121,\ 16,\ 49,\ 169,\ 256,\ 169,\ \underline{\ \ }$$

We have already seen the term 169. So, we know the next term is 256.

$$65,\ 121,\ 16,\ 49,\ 169,\ 256,\ 169,\ 256$$

148. The first term is 65. The squares of the digits of 65 are $6^2=36$ and $5^2=25$. So, the next term is $36+25=61$.

$$65,\ 61,\ \underline{\ \ },\ \underline{\ \ },\ \underline{\ \ },\ \underline{\ \ },\ \underline{\ \ },\ \underline{\ \ }$$

The squares of the digits of 61 are $6^2=36$ and $1^2=1$. So, the next term is $36+1=37$.

$$65,\ 61,\ 37,\ \underline{\ \ },\ \underline{\ \ },\ \underline{\ \ },\ \underline{\ \ },\ \underline{\ \ }$$

The next term is $3^2+7^2=9+49=58$.

$$65,\ 61,\ 37,\ 58,\ \underline{\ \ },\ \underline{\ \ },\ \underline{\ \ },\ \underline{\ \ }$$

The next term is $5^2+8^2=25+64=89$.

$$65,\ 61,\ 37,\ 58,\ 89,\ \underline{\ \ },\ \underline{\ \ },\ \underline{\ \ }$$

The next term is $8^2+9^2=64+81=145$.

$$65,\ 61,\ 37,\ 58,\ 89,\ 145,\ \underline{\ \ },\ \underline{\ \ }$$

The next term is $1^2+4^2+5^2=1+16+25=42$.

$$65,\ 61,\ 37,\ 58,\ 89,\ 145,\ 42,\ \underline{\ \ }$$

The next term is $4^2+2^2=16+4=20$.

$$65,\ 61,\ 37,\ 58,\ 89,\ 145,\ 42,\ 20$$

If we continue this sequence, does it ever repeat?

149. The first two terms of the sequence are 0 and 1. Since each term is the sum of all previous terms, the 3rd term is $0+1=1$.

The 4th term is $0+1+1=2$.
The 5th term is $0+1+1+2=4$.
The 6th term is $0+1+1+2+4=8$.
The 7th term is $0+1+1+2+4+8=16$.

After the first two terms, each term in the sequence is a power of 2:

1st	2nd	3rd	4th	5th	6th	7th	...
0,	1,	1,	2,	4,	8,	16,	...

$$0,\ 1,\ 2^0,\ 2^1,\ 2^2,\ 2^3,\ 2^4,\ ...$$

The exponent in the power of 2 is three less than the term's position number in the sequence. For example, the 5th term is $2^{5-3}=2^2$.

So, the nth term in the sequence is $\mathbf{2^{(n-3)}}$ for values of n that are greater than 2 (we make this distinction since $2^{(n-3)}$ works for all terms except the 1st and 2nd).

150. a. To find the term that comes after 32, we reverse the digits of 32 and add 1 to the result: $23+1=24$.

Then, the next term is $42+1=43$.
Then, the next term is $34+1=35$.

We continue this pattern to complete the sequence as shown.

$$32,\ 24,\ 43,\ 35,\ 54,\ 46,\ 65,\ 57,\ 76,\ 68$$

b. We work backwards. The final term is 100. So, reversing the previous term's digits and adding 1 to the result gives 100. Therefore, the reverse of the previous term's digits is $100 - 1 = 99$. Since the reverse of 99's digits is 99, the term before 100 is 99.

$$\underline{}, \underline{}, \underline{}, \underline{}, \underline{}, \underline{}, \underline{}, \underline{}, \mathbf{99}, 100$$

To find the term before 99, we subtract $99 - 1 = 98$, then reverse the digits to get 89.

$$\underline{}, \underline{}, \underline{}, \underline{}, \underline{}, \underline{}, \underline{}, \mathbf{89}, \mathbf{99}, 100$$

To find the term before 89, we subtract $89 - 1 = 88$, then reverse the digits to get 88.

$$\underline{}, \underline{}, \underline{}, \underline{}, \underline{}, \underline{}, \mathbf{88}, \mathbf{89}, \mathbf{99}, 100$$

We continue working from right to left to complete the sequence as shown below.

55, **56**, **66**, **67**, **77**, **78**, **88**, **89**, **99**, 100

151. a. The term that comes after 85 and 52 is $85 - 52 = 33$.

The term after 52 and 33 is $52 - 33 = 19$.
The term after 33 and 19 is $33 - 19 = 14$.
The term after 19 and 14 is $19 - 14 = 5$.
The term after 14 and 5 is $14 - 5 = 9$.
The term after 5 and 9 is $9 - 5 = 4$.
The term after 9 and 4 is $9 - 4 = 5$.
The term after 4 and 5 is $5 - 4 = 1$.

The completed sequence is shown below.

85, 52, **33**, **19**, **14**, **5**, **9**, **4**, **5**, **1**

b. The term that comes after 76 and 47 is $76 - 47 = 29$.

The term after 47 and 29 is $47 - 29 = 18$.
The term after 29 and 18 is $29 - 18 = 11$.
The term after 18 and 11 is $18 - 11 = 7$.
The term after 11 and 7 is $11 - 7 = 4$.
The term after 7 and 4 is $7 - 4 = 3$.
The term after 4 and 3 is $4 - 3 = 1$.
The term after 3 and 1 is $3 - 1 = 2$.

The completed sequence is shown below.

76, 47, **29**, **18**, **11**, **7**, **4**, **3**, **1**, **2**

152. The term after 5 and 3 is $5 - 3 = 2$, giving 5, 3, 2, ...

Next, we have $3 - 2 = 1$, giving 5, 3, 2, 1, ...
Next, we have $2 - 1 = 1$, giving 5, 3, 2, 1, 1, ...
Next, we have $1 - 1 = 0$, giving 5, 3, 2, 1, 1, 0, ...
Next, we have $1 - 0 = 1$, giving 5, 3, 2, 1, 1, 0, 1, ...
Next, we have $1 - 0 = 1$, giving 5, 3, 2, 1, 1, 0, 1, 1, ...
Next, we have $1 - 1 = 0$, giving 5, 3, 2, 1, 1, 0, 1, 1, 0, ...

Beginning with the 4th term, we have a repeating cycle of the terms 1, 1, 0.

The terms equal to 0 are the 6th, 9th, 12th, 15th, and so on. These are the terms whose position number is a multiple of 3, starting with the 6th term.

Since 75 is a multiple of 3, the 75th term is **0**.

153. There are many correct answers to this problem. The key is working backwards! We begin by writing two numbers for the last two terms, where the next-to-last term is greater than the last term. For example, suppose the last two terms are 2 and 1. We let n represent the term before 2.

$$\underline{}, \underline{}, \underline{}, \underline{}, \underline{}, \underline{}, \underline{}, n, 2, 1$$

There are two possible values for n:

- If n is less than 2, then $2 - n = 1$, giving $n = 1$.
- If n is greater than 2, then $n - 2 = 1$, giving $n = 3$.

Since our terms decrease from left to right, n must be greater than 2. So, we compute n by adding $1 + 2 = 3$.

$$\underline{}, \underline{}, \underline{}, \underline{}, \underline{}, \underline{}, \underline{}, 3, 2, 1$$

Working from right to left, we continue adding the two largest terms to get the next term. For example, the term to the left of 3 is $2 + 3 = 5$, followed by $3 + 5 = 8$, and so on. Since the terms *increase* from right to left, they *decrease* from left to right, as shown below.

89, 55, 34, 21, 13, 8, 5, 3, 2, 1

Notice that these are the Fibonacci numbers in reverse! You can use numbers other than 2 and 1 to end your sequence to get a different answer. You can check your work by adding each pair of numbers from right to left!

154. The sum of the digits of a two-digit number is at least $1 + 0 = 1$ and at most $9 + 9 = 18$. So, the 2nd term in Ralph's sequence is the product of 9 and some number between 1 and 18, which gives the following possibilities for the 2nd term:

9, 18, 27, ..., 144, 153, 162.

For any number divisible by 9, the sum of that number's digits is also divisible by 9. Therefore, since the 2nd term is a multiple of 9, the sum of the digits of the 2nd term is also a multiple of 9.

Also, since the 2nd term has three or fewer digits, and the 2nd term cannot be 999, the sum of the digits of the 2nd term is less than $9 + 9 + 9 = 27$. So, the sum of the digits of the 2nd term is 9 or 18.

Therefore, the 3rd term in Ralph's sequence is $9 \cdot 9 = 81$ or $9 \cdot 18 = 162$.

The sum of the digits of both 81 and 162 is 9. So, the 4th term must be $9 \cdot 9 = 81$. Therefore, there is only **1** number that could be the 4th term in Ralph's sequence.

SEQUENCES
Challenge Problems   32–35

155. We look for a pattern in the units digit of the first few powers of 3. Since we are only concerned with the units digit, we can ignore the other digits in each power. *(Review units digits in the Multiplication chapter of Beast Academy 4A.)*

$3^1 = \underline{3}$.
$3^2 = \underline{9}$.
$3^3 = 2\underline{7}$.
$3^4 = 3^3 \cdot 3$ has the same units digit as $7 \cdot 3 = 2\underline{1}$.
$3^5 = 3^4 \cdot 3$ has the same units digit as $1 \cdot 3 = \underline{3}$.
$3^6 = 3^5 \cdot 3$ has the same units digit as $3 \cdot 3 = \underline{9}$.
$3^7 = 3^6 \cdot 3$ has the same units digit as $9 \cdot 3 = 2\underline{7}$.
$3^8 = 3^7 \cdot 3$ has the same units digit as $7 \cdot 3 = 2\underline{1}$.

The units digits follow the repeating pattern

$$3, 9, 7, 1, 3, 9, 7, 1, \ldots .$$

These digits repeat every 4 terms. The terms in this pattern equal to 1 are the 4th, 8th, 12th, 16th, and so on. These are the terms whose position number is a multiple of 4. Since $96 = 4(24)$, the 96th term in this pattern is 1.

Therefore, 3^{96} has units digit 1. Continuing the pattern, 3^{97} has units digit 3, 3^{98} has units digit 9, and 3^{99} has units digit **7**.

156. Rather than computing the entire sum, we consider the sums of the first few terms and look for a pattern.

The sum of the first 2 terms is $1 + 2 = 3$.
The sum of the first 3 terms is $1 + 2 + 4 = 7$.
The sum of the first 4 terms is $1 + 2 + 4 + 8 = 15$.
The sum of the first 5 terms is $1 + 2 + 4 + 8 + 16 = 31$.

We notice that each result is one less than the next power of 2 in the sum! For example, the sum of the first 5 terms is 31, which is one less than the next term in the sum, $2^5 = 32$.

$$\underbrace{1 + 2 + 4 + 8 + 16}_{31} + 32 + \cdots$$

Continuing this pattern, the sum of the first 6 terms is $31 + 32 = 63$, which is one less than the next term in the sum, $2^6 = 64$.

So, the sum of all of the terms is one less than the power of two after $2^{12} = 4{,}096$. The power of two after 4,096 is $2^{13} = 2 \cdot 4{,}096 = 8{,}192$. So, the sum of all the terms in the sum is $8{,}192 - 1 = \textbf{8,191}$.

In general, the sum of the first n powers of 2 (starting with $2^0 = 1$) is $2^n - 1$.

157. In this pattern we move through the positive integers, starting with 1 in the top-left corner and following the path as shown.

```
A   B   C   D   E   F
1 — 2 — 3 — 4 — 5 — 6
12 11 10  9   8   7
13 14 15 16  17  18
24 23 22 21  20  19
 :   :   :   :   :   :
```

This pattern of moving right, then down, then left, then down repeats every 12 integers. So, we can add 12 to any number and end up in the same column we started in.

For example, 1 is in column A, and $1 + 12 = 13$ is also in column A. Similarly, 8 is in column E, and $8 + 12 = 20$ is also in column E.

So, to find the column 99 is under, we can continually subtract 12 from 99 until we arrive at a number whose column we know. Since $99 - 12(8) = 99 - 96 = 3$, we know 99 is under the same column as 3, which is column **C**.

— or —

The pattern of moving right, then down, then left, then down repeats every 12 integers. So, every multiple of 12 is in the same column and occurs at the same point in the cycle. So, $12 \cdot 8 = 96$ is in column A, and the path moves to the right after passing through 96.

We continue the path from 96 until we reach 99, which is in column **C**.

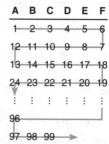

158. This spiraling pattern forms a roughly square-shaped figure. This suggests that thinking in terms of squares might be useful. We consider the various square pieces of the pattern, as shown below.

```
                                              21—22—23—24—25
          7—8—9       7—8—9—10          20  7—8—9—10
  1—2     6   1—2     6   1—2  11        19  6   1—2  11
  4—3     5—4—3       5—4—3   12         18  5—4—3   12
                      16—15—14—13        17—16—15—14—13
```

The last number in each figure above is a perfect square.

• $2^2 = 4$ is at the lower-left corner of the 2-by-2 square.

• $3^2 = 9$ is at the upper-right corner of the 3-by-3 square.

• $4^2 = 16$ is at the lower-left corner of the 4-by-4 square.

• $5^2 = 25$ is at the upper-right corner of the 5-by-5 square.

Each square of an even integer is in a lower-left corner. Each square of an odd integer is in an upper-right corner.

Since $100 = 10^2$, the number 100 will be in the lower-left corner of a 10-by-10 square. To determine what number will be directly above 100, we look at a known example.

```
  21—22—23—24— ···
  20  7—8—9—10
  19  6   1—2  11
  18  5—④—3   12
  17—⑯—15—14—13
```

Here, we see that the number directly above $4^2 = 16$ is the number to the left of the previous square of an even integer, $2^2 = 4$.

Similarly, the number directly above $10^2 = 100$ is the number to the left of the previous square of an even integer, $8^2 = 64$.

The number to the left of 64 is 65. So, the number directly above 100 is **65**.

159. We compute the remainder when dividing the first few powers of 2 by 7 and look for a pattern.

The remainder of $2^1 \div 7 = 2 \div 7$ is 2.
The remainder of $2^2 \div 7 = 4 \div 7$ is 4.
The remainder of $2^3 \div 7 = 8 \div 7$ is 1.
The remainder of $2^4 \div 7 = 16 \div 7$ is 2.
The remainder of $2^5 \div 7 = 32 \div 7$ is 4.
The remainder of $2^6 \div 7 = 64 \div 7$ is 1.

We see a pattern! Since each power of 2 is 2 times the power that came before it, each remainder is the remainder of

(2 times the remainder that came before) $\div 7$.

So, we can be sure that this pattern continues. *(Review this remainder concept in the Division chapter of Beast Academy 3C.)*

So, the remainders follow the repeating pattern

2, 4, 1, 2, 4, 1, 2, 4, 1, ...

The terms equal to 1 are the 3rd, 6th, 9th, 12th, and so on. These are the terms whose position number is a multiple of 3. Since $99 = 3(33)$, the 99th term in this pattern is 1.

So, the remainder of $2^{99} \div 7$ is 1. Continuing the pattern, the remainder of $2^{100} \div 7$ is **2**.

160. The compubot displays the number 3. Since 3 is odd, pressing the ★ button gives the number $3(3) + 1 = 10$.

10 is even, so pressing ★ again gives $10 \div 2 = 5$.
5 is odd, so pressing ★ again gives $3(5) + 1 = 16$.
16 is even, so pressing ★ again gives $16 \div 2 = 8$.
8 is even, so pressing ★ again gives $8 \div 2 = 4$.
4 is even, so pressing ★ again gives $4 \div 2 = 2$.
2 is even, so pressing ★ again gives $2 \div 2 = 1$.
1 is odd, so pressing ★ again gives $3(1) + 1 = 4$.

We have already seen the number 4, so we know the next number will be 2, followed by 1, then back to 4, and so on forever.

We can summarize our findings with the chart below.

Starting number: 3

★ Presses	1	2	3	4	5	6	7	8	9	10
Result	10	5	16	8	4	2	1	4	2	1

Starting with the 5th time the ★ is pressed, the results repeat every 3 presses. So, 2 is displayed after the 6th press, 9th press, 12th press, 15th press, and so on. In general, if the number of times the ★ has been pressed is a multiple of 3 (and greater than 5), then the number displayed will be 2.

Since $75 = 3(25)$, the number displayed after the 75th press is **2**.

This question is related to a very famous problem. Many mathematicians suspect that no matter what positive integer you start with, you will eventually reach the number 1 (in this case, we started with 3 and reached 1 after the 7th press). This is known as the Collatz Conjecture, and mathematicians are still trying to prove that it is true!

161. We consider how much liquid is in each juice box after the first several pours. We'll call the first juice box Box A and the second juice box Box B.

1st Pour
Grogg pours half of the 2 cups of juice in Box A into Box B.

Half of 2 is $\frac{1}{2} \cdot 2 = 1$.

So, Box A is left with $2 - 1 = 1$ cup, and Box B is left with $0 + 1 = 1$ cup.

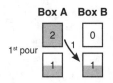

2nd Pour
Grogg pours one third of the 1 cup of juice in Box B into Box A.

One third of 1 is $\frac{1}{3} \cdot 1 = \frac{1}{3}$.

So, Box B is left with $1 - \frac{1}{3} = \frac{2}{3}$ cups, and Box A is left with $1 + \frac{1}{3} = \frac{4}{3}$ cups.

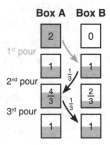

3rd Pour
Grogg pours one fourth of the $\frac{4}{3}$ cups of juice in Box A into Box B.

One fourth of $\frac{4}{3}$ is $\frac{1}{4} \cdot \frac{4}{3} = \frac{1}{3}$.

So, Box A is left with $\frac{4}{3} - \frac{1}{3} = 1$ cup, and Box B is left with $\frac{2}{3} + \frac{1}{3} = 1$ cup.

4th Pour
Grogg pours one fifth of the 1 cup of juice in Box B into Box A.

One fifth of 1 is $\frac{1}{5} \cdot 1 = \frac{1}{5}$.

So, Box B is left with $1 - \frac{1}{5} = \frac{4}{5}$ cups, and Box A is left with $1 + \frac{1}{5} = \frac{6}{5}$ cups.

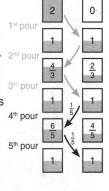

5th Pour
Grogg pours one sixth of the $\frac{6}{5}$ cups of juice in Box A into Box B.

One sixth of $\frac{6}{5}$ is $\frac{1}{6} \cdot \frac{6}{5} = \frac{1}{5}$.

So, Box A is left with $\frac{6}{5} - \frac{1}{5} = 1$ cup, and Box B is left with $\frac{4}{5} + \frac{1}{5} = 1$ cup.

After the first pour, each pair of consecutive pours shown above moved the same amount of juice out of and then back into Box B. This left both boxes with 1 cup of juice after the odd-numbered pours. We explore whether or not this pattern will continue.

When both boxes have 1 cup of juice, Grogg's next pour moves some unit fraction of Box B's juice into Box A. If we call that unit fraction $\frac{1}{n}$, then Grogg pours $\frac{1}{n} \cdot 1 = \frac{1}{n}$ cups from Box B into Box A.

This leaves Box A with $1 + \frac{1}{n}$ cups of juice. Since $1 = \frac{n}{n}$, we can rewrite $1 + \frac{1}{n}$ as shown below:

$$1 + \frac{1}{n} = \frac{n}{n} + \frac{1}{n} = \frac{n+1}{n}.$$

So, Box A has $\frac{n+1}{n}$ cups of juice.

Grogg's next pour moves a unit fraction of Box A's juice into Box B. The denominator of this unit fraction is one greater than on the previous pour. So, Grogg pours $\frac{1}{n+1}$ of the $\frac{n+1}{n}$ cups in Box A into Box B, which is

$$\frac{1}{n+1}\cdot\frac{n+1}{n}=\frac{1}{n}\text{ cups.}$$

This is the same amount that Grogg moved *out* of Box B from the previous pour. So, after both pours, Box B has the same amount of juice it started with: 1 cup.

So, the amount of juice in each box is always 1 cup after each odd-numbered pour. Therefore, there is **1** cup of juice in the first box after Grogg's 99th pour.

162. Since the common difference is 8, we can continually add 8 to -999 until we get a positive result. $8(125) = 1,000$, so $-999+8(125) = -999+1,000 = 1$ is a term in the sequence.

There is no smaller positive integer than 1, so the smallest positive term that appears in the sequence is **1**.

163. This is the sequence of positive integers written in base-2. *(Review base-2 numbers in the Exponents chapter of Beast Academy 4A.)*

Alternatively, we can think of it as the sequence of positive integers that can be written using only the digits 0 and 1, ordered from least to greatest.

After 111, the next-smallest number is 1000, followed by 1001, then 1010, then 1011, then 1100.

1, 10, 11, 100, 101, 110, 111, **1000, 1001, 1010, 1011, 1100**

164. In the list of positive integers where the perfect squares *are not* removed, the 50th term is 50.

There are 7 perfect squares less than 50:

1, 4, 9, 16, 25, 36, and 49.

So, in the sequence where the perfect squares *are* removed, all 7 of these terms have been taken out. So, 50 appears in position number $50-7 = 43$. We count up to the 50th term from here.

Position: 43rd 44th 45th 46th 47th 48th 49th 50th
Term: 50, 51, 52, 53, 54, 55, 56, 57

There are no perfect squares from 50 to 57 to remove, so the 50th term is **57**.

165. Since we begin with 0, and adding 1 then 2 then 3 is the same as adding 6, we know that every multiple of 6 is a term in this sequence.

The smallest 7-digit number is 1,000,000. We look for a multiple of 6 that is close to 1,000,000. Since 1,000,002 is even and the sum of its digits is divisible by 3, it is divisible by both 2 and by 3, and is therefore a multiple of 6. So, 1,000,002 is a term in the sequence.

Each multiple of 6 in the sequence is reached after adding 3 to the previous term. So, the term that came before 1,000,002 is $1,000,002-3 = 999,999$. However, 999,999 is a 6-digit number.

So, the smallest 7-digit number that appears in the sequence is **1,000,002**.

166. The 7 terms in the sequence have sum 133, so their average is $133\div7 = 19$. In an arithmetic sequence, the average of the terms is equal to the median. So, the median is 19.

___, ___, ___, 19, ___, ___, ___

We also know that the first and last terms in an arithmetic sequence balance around the median. So, the first and last terms balance around 19.

To make the last term as great as possible, we make the first term as small as possible. Since every term is a positive integer, the smallest possible value of the first term is 1.

Then, since 1 is 18 less than the median, the last term must be 18 more than the median: $19+18 = 37$.

$$\overset{-18}{\overbrace{}}\quad\overset{+18}{\overbrace{}}$$
1, ___, ___, 19, ___, ___, 37

We are told that each term is a positive integer, so we must check that there are no non-integer terms in this sequence. The 1st term is 1 and the 4th term is 19, so the common difference is $\frac{19-1}{4-1} = \frac{18}{3} = 6$. Since the common difference is an integer, and the first term is an integer, each term in the sequence is also an integer.

So, the greatest possible term that can appear in the sequence is **37**.

We could have also reversed the order of the terms in this sequence to arrive at the same final answer.

167. Since each term is the sum of the three previous terms, we know that $x+y+z = 44$.

We also know that $y+z+44 = 81$. We can subtract 44 from both sides of this equation to get $y+z = 37$.

In the equation $x+y+z = 44$, we see the expression $y+z$. We know from our work above that $y+z$ is 37, so we replace $y+z$ with 37 in this equation.

$$x+y+z = 44,$$
$$x + 37 = 44.$$

Now we have an equation with just one variable! Solving, we have $x = \mathbf{7}$.

As an extra challenge, can you determine possible values for y and z? Are these the only possible values?

168. The 1st figure uses 3 toothpicks.
The 2nd figure uses $3+6 = 9$ toothpicks.
The 3rd figure uses $3+6+9 = 18$ toothpicks.

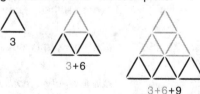

Continuing the pattern, the nth figure uses $3+6+9+\cdots+3n$ toothpicks. We can factor a 3 out of each term to get

$$3+6+9+\cdots+3n = 3(1+2+3+\cdots+n).$$

We learned previously that $1+2+3+\cdots+n = \frac{n(n+1)}{2}$. So,

$$3(1+2+3+\cdots+n) = 3\cdot\frac{n(n+1)}{2}.$$

So, the number of toothpicks needed to make the 20th figure in the pattern is $3 \cdot \frac{20(21)}{2} = 3 \cdot 210 = \textbf{630}$.

— *or* —

It takes 3 toothpicks to make each small triangle. However, when triangles in a figure touch, some or all of their toothpicks are shared by another triangle. This makes counting the number of toothpicks based on the number of triangles difficult.

However, if we count just the triangles "pointing up," we count every toothpick in the figure, and there are no shared toothpicks. In the figure below, the "up" triangles are shown with a dot in the center.

There are 6 "up" triangles, each with 3 toothpicks, for a total of $6 \cdot 3 = 18$ toothpicks.

So, we want to know the number of "up" triangles that are in the 20th figure in this pattern. Looking at the first three figures, we have 1 "up" triangle, then 3, then 6.

The "up" triangles make a pattern of triangular numbers! So, the number of "up" triangles in the 20th figure is the 20th triangular number: $\frac{20 \cdot 21}{2} = 10 \cdot 21 = 210$. Each "up" triangle uses 3 toothpicks, so the number of toothpicks in the 20th figure of this pattern is $3 \cdot 210 = \textbf{630}$.

169. Bronkle writes an arithmetic sequence with first term 40 and common difference 6. So, the kth term of Bronkle's sequence is $40 + (k-1)6 = 40 + 6k - 6 = 6k + 34$.

Gergum writes an arithmetic sequence with first term -50 and common difference 9. So, the kth term of Gergum's sequence is $-50 + (k-1)9 = -50 + 9k - 9 = 9k - 59$.

We want to know when the kth term of Bronkle's sequence is equal to the kth term of Gergum's sequence. So, we write an equation:

$$6k + 34 = 9k - 59.$$

To isolate k, we begin by subtracting $6k$ from both sides of the equation. This gives

$$34 = 3k - 59.$$

Then, adding 59 to both sides gives $93 = 3k$. Dividing both sides by 3, we have $31 = k$.

So, the 31st terms of Bronkle's and Gergum's sequences are equal. We can replace k with 31 in either $6k + 34$ or $9k - 59$ to find the value of the 31st term:

$$6k + 34 = 6(31) + 34 = 186 + 34 = 220.$$
$$9k - 59 = 9(31) - 59 = 279 - 59 = 220.$$

So, $k = 31$, and the 31st term of each sequence is **220**.

RATIOS & RATES
Ratios   37–39

1. There are 3 black circles and 7 white circles, so the ratio of black circles to white circles is **3 to 7**.

2. There are 10 circles and 7 triangles, so the ratio of circles to triangles is **10 to 7**.

3. There are 5 black shapes and 12 white shapes, so the ratio of black shapes to white shapes is **5 to 12**.

4. There are 5 white triangles and 2 black triangles, so the ratio of white triangles to black triangles is **5 to 2**.

5. We can split these 12 circles into groups of 3 circles, then shade 2 circles in each group. One example is shown below.

All together, we must shade **8** circles.

6. We divide 5 and 10 by their greatest common factor, 5, to get $5{:}10 = 1{:}2$. The greatest common factor of 1 and 2 is 1, so **1:2** is the simplest form of this ratio.

7. We divide 197 and 197 by their greatest common factor, 197, to get $197{:}197 = $ **1:1**.

8. The greatest common factor of 25 and 9 is 1, so **25:9** is the simplest form of this ratio.

9. We divide 40 and 16 by their greatest common factor, 8, to get $40{:}16 = $ **5:2**.

10. We divide 42 and 70 by their greatest common factor, 14, to get $42{:}70 = $ **3:5**.

11. We divide 121 and 88 by their greatest common factor, 11, to get $121{:}88 = $ **11:8**.

12. We divide 24 and 92 by their greatest common factor, 4, to get $24{:}92 = $ **6:23**.

13. We divide 91 and 65 by their greatest common factor, 13, to get $91{:}65 = $ **7:5**.

14. $2\frac{1}{3} = \frac{7}{3}$. To write $2\frac{1}{3}{:}7 = \frac{7}{3}{:}7$ as a ratio of two integers, we begin by multiplying both quantities by 3 to eliminate the fraction. Then, we have $\frac{7}{3}{:}7 = 7{:}21$. We divide each of 7 and 21 by their greatest common factor, 7, to get $7{:}21 = 1{:}3$. Therefore, $2\frac{1}{3}{:}7 = $ **1:3**.

15. To write $\frac{1}{3}{:}\frac{3}{5}$ as a ratio of two integers, we multiply $\frac{1}{3}$ and $\frac{3}{5}$ by the least common multiple of their denominators, $3 \cdot 5 = 15$. This gives $\frac{1}{3}{:}\frac{3}{5} = $ **5:9**.

16. The ratio of blue cars to red cars is 16:10, which simplifies to **8:5**.

17. The class has 26 students with brown eyes and $30 - 26 = 4$ students with green eyes. So, the ratio of brown-eyed students to green-eyed students is $26{:}4 = $ **13:2**.

18. There were 56 flips that landed heads, so $100 - 56 = 44$ of the flips landed tails. The ratio of heads flipped to tails flipped is $56{:}44 = $ **14:11**.

19. We can divide the smaller square into $3 \cdot 3 = 9$ little squares, as shown below. Since the ratio of the side length of the smaller square to the side length of the larger square is 3:5, we can place 5 of the same little squares along each side of the larger square. So, we can divide the larger square into $5 \cdot 5 = 25$ little squares.

Therefore, the ratio of the area of the smaller square to the area of the larger square is **9:25**.

20. The ratio of cups of yogurt to cups of strawberries is $\frac{2}{3}{:}\frac{1}{2}$. To write this as a ratio of two integers, we multiply $\frac{2}{3}$ and $\frac{1}{2}$ by the least common multiple of their denominators, $3 \cdot 2 = 6$. This gives $\frac{2}{3}{:}\frac{1}{2} = $ **4:3**.

RATIOS & RATES
Using Ratios   40–41

21. We can split the dancers into groups, each with 5 dragons and 3 yetis. Since there are 30 dragons, we can make $30 \div 5 = 6$ groups with 5 dragons in each group. Then, each of the 6 groups also has 3 yetis. So, there are $6 \cdot 3 = $ **18** yetis in the class.

22. We can split fans at the tournament into groups, each with 5 Beast Academy fans and 4 Orb Academy fans. Since there are 160 Beast Academy fans, we can make $160 \div 5 = 32$ groups with 5 Beast Academy fans in each group. Each of the 32 groups also has 4 Orb Academy fans. So, there are $32 \cdot 4 = $ **128** Orb Academy fans.

23. We can split the sodas sold into groups, each with 7 grape and 6 orange sodas. Since 84 orange sodas were sold, we can make $84 \div 6 = 14$ groups with 6 orange sodas in each group. Then, each of the 14 groups also has 7 grape sodas. So, $14 \cdot 7 = $ **98** grape sodas were sold.

24. We can split Priti's answers into groups, each with 2 incorrect answers and 5 correct answers. Since she answered 30 problems correctly, we can make $30 \div 5 = 6$ groups with 5 correct answers in each group. Then, each of the 6 groups also has 2 incorrect answers. So, Priti answered $6 \cdot 2 = 12$ questions incorrectly, and there were a total of $30 + 12 = $ **42** questions on the test.

— *or* —

As in the previous approach, we can make $30 \div 5 = 6$ groups with 5 correct answers in each group. Each of the 6 groups contains a total of $2 + 5 = 7$ answers. So, there were $6 \cdot 7 = $ **42** questions on the test.

25. We can split Roland's cookies into groups, each with 4 sugar and 3 oatmeal-raisin. Since Roland baked 36 oatmeal-raisin cookies, we can make $36 \div 3 = 12$ groups with 3 oatmeal-raisin cookies each. Then, each of the 12 groups also has 4 sugar cookies. So, Roland baked $12 \cdot 4 = 48$ sugar cookies, which makes a total of $36 + 48 = \mathbf{84}$ cookies.

— *or* —

As in the previous approach, we can make $36 \div 3 = 12$ groups with 3 oatmeal-raisin cookies each. Since each of the groups contains $3 + 4 = 7$ cookies total, Roland baked $7 \cdot 12 = \mathbf{84}$ cookies all together.

26. We can split Allison's green paint mix into 5 equal parts: 2 parts blue paint and 3 parts yellow paint. Since the 3 parts of yellow paint equal $\frac{1}{4}$ of a pint, each part is $\frac{1}{4} \div 3 = \frac{1}{4} \cdot \frac{1}{3} = \frac{1}{12}$ of a pint. Therefore, the 2 parts of green paint equal $2 \cdot \frac{1}{12} = \frac{2}{12} = \frac{1}{6}$ of a pint.

Check: The ratio of blue to yellow paint is $\frac{1}{6} : \frac{1}{4}$. Multiplying both numbers by the least common multiple of their denominators, 12, gives $\frac{1}{6} : \frac{1}{4} = 2 : 3$. ✓

27. We can split Peter's miles on a bus or train into 10 equal parts: 7 parts train-miles and 3 parts bus-miles. Since the 3 parts of bus-miles equal 210 miles, each part is $210 \div 3 = 70$ miles. So, the 7 parts of train-miles equal $7 \cdot 70 = 490$ miles, and Peter traveled a total of $210 + 490 = 700$ miles by bus or train.

Then, Peter biked 1 mile for every 14 miles that he rode the bus or train. Since he traveled 700 miles by bus or train and $700 \div 14 = 50$, he biked **50** miles.

28. For every 3 white beads that Anna uses, she'll use 7 black beads.

Since $39 = 3 \cdot 13$ is the largest multiple of 3 that is less than 40, Anna can make up to 13 groups of 3 white beads each (with one left over).

Since $84 = 7 \cdot 12$ is the largest multiple of 7 that is less than 90, Anna can make up to 12 groups of 7 black beads each (with six left over).

So, when making groups of 7 black *and* 3 white beads, Anna is limited by the black beads she has. So, she cannot make more than 12 groups of 7 black and 3 white beads. Anna can use at most $3 \cdot 12 = \mathbf{36}$ white beads.

RATIOS & RATES
The Whole Amount   42-43

29. Each blorble has $3 + 4 = 7$ horns. So, there are $98 \div 7 = 14$ blorbles on the field. Fourteen blorbles have $14 \cdot 3 = 42$ long horns and $14 \cdot 4 = \mathbf{56}$ short horns.

30. For every 4 players wearing white jerseys, there are 5 players wearing blue jerseys. We can split 27 players into $27 \div 9 = 3$ groups of 9 players, each group having 4 players wearing white and 5 players wearing blue.

So, $3 \cdot \underline{4} = \mathbf{12}$ players are wearing white jerseys and $3 \cdot \underline{5} = 15$ are wearing blue.

— *or* —

For every 4 equal groups of players wearing white, there are 5 equal groups of players wearing blue.

We can split 27 players into $4 + 5 = 9$ equal groups, each with $27 \div 9 = 3$ players.

So, $3 \cdot \underline{4} = \mathbf{12}$ players are wearing white jerseys and $3 \cdot \underline{5} = 15$ are wearing blue.

— *or* —

$\frac{4}{9}$ of the jerseys are white, and $\frac{5}{9}$ are blue. So, at a game with 27 players, $\frac{4}{9} \cdot 27 = \mathbf{12}$ are wearing white and $\frac{5}{9} \cdot 27 = 15$ are wearing blue.

31. For every 3 defenders, there are 5 strikers. We can split 96 players into $96 \div 8 = 12$ teams of 8 players, each with 3 defenders and 5 strikers.

So, $12 \cdot \underline{3} = \mathbf{36}$ of the players are defenders and $12 \cdot \underline{5} = 60$ are strikers.

— *or* —

$\frac{3}{8}$ of the players are defenders, and $\frac{5}{8}$ are strikers. So, in a league of 96 players, $\frac{3}{8} \cdot 96 = \mathbf{36}$ are defenders and $\frac{5}{8} \cdot 96 = 60$ are strikers.

32. On every tray of 10 cupcakes, 4 are chocolate and 6 are vanilla. We can split 240 cupcakes into $240 \div 10 = 24$ trays of 10 cupcakes, each with 4 chocolate and 6 vanilla. So, $24 \cdot \underline{4} = 96$ of the cupcakes are chocolate and $24 \cdot \underline{6} = \mathbf{144}$ are vanilla.

— *or* —

$\frac{4}{10} = \frac{2}{5}$ of the cupcakes are chocolate, and $\frac{6}{10} = \frac{3}{5}$ are vanilla. So, of the 240 cupcakes, $\frac{2}{5} \cdot 240 = 96$ are chocolate and $\frac{3}{5} \cdot 240 = \mathbf{144}$ are vanilla.

33. Since the ratio of stingravens to pandakeets is 2:9, $\frac{2}{11}$ of the 176 animals in the exhibit are stingravens. $\frac{2}{11} \cdot 176 = 32$, so there are 32 stingravens.

Since the ratio of spotted to striped stingravens is 3:5, $\frac{5}{8}$ of the 32 stingravens are striped. $\frac{5}{8} \cdot 32 = 20$, so there are **20** striped stingravens in the exhibit.

34. The width-to-height ratio of 5:2 tells us that $\frac{5}{7}$ of the perimeter comes from the width. So, $\frac{5}{7} \cdot 210 = 150$ inches of the perimeter come from the width, and the remaining $210 - 150 = 60$ inches come from the height.

The perimeter of a rectangle includes two copies each of the width and height. So, the rectangle is $150 \div 2 = 75$ inches wide and $60 \div 2 = 30$ inches tall. The area of a 75-by-30-inch rectangle is $75 \cdot 30 = \mathbf{2{,}250}$ square inches.

— *or* —

For every 5 inches in the width of the rectangle, there are 2 inches in the height. So, for some value of x, the rectangle is $5x$ inches wide and $2x$ inches tall. Then, the perimeter of the rectangle is $5x + 5x + 2x + 2x = 14x$ inches.

Since $14x = 210$, we have $x = 15$.

So, the rectangle is $5 \cdot 15 = 75$ inches wide and $2 \cdot 15 = 30$ inches tall. The area of a 75-by-30-inch rectangle is $75 \cdot 30 = \mathbf{2{,}250}$ square inches.

35. Simplifying each ratio, we get

$21:28 = 3:4,$ $63:84 = 3:4,$ $39:52 = 3:4,$
$24:36 = 2:3,$ $51:68 = 3:4.$

All but $24:36 = 2:3$ are equivalent to $3:4$. So, $24:36$ is the ratio that is not equivalent to the other four.

21:28   63:84   39:52   (24:36)   51:68

36. The ratio $11:17$ is already in simplest form. A ratio of integers whose left-side quantity is not a multiple of 11 cannot be simplified to

11:(some integer).

So, we can eliminate the two ratios crossed out below.

22:34   ~~30:48~~   33:51   44:68
55:85   55:88   ~~60:93~~   66:102

Simplifying the remaining ratios, we have

$22:34 = 11:17,$ $33:51 = 11:17,$ $44:68 = 11:17,$
$55:85 = 11:17,$ $55:88 = 5:8,$ $66:102 = 11:17.$

So, the five ratios circled below are equivalent to $11:17$.

(22:34)   30:48   (33:51)   (44:68)
(55:85)   55:88   60:93   (66:102)

37. Simplifying each ratio, we get

$21:35 = 3:5,$ $14:24 = 7:12,$ $32:48 = 2:3,$
$56:96 = 7:12,$ $55:90 = 11:18,$ $33:54 = 11:18,$
$12:20 = 3:5,$ $18:27 = 2:3.$

So, the pairs of equivalent ratios are

$21:35 = 12:20,$ $14:24 = 56:96,$
$32:48 = 18:27,$ and $55:90 = 33:54.$

You may have written these equations in a different order.

38. Since we want three equivalent ratios, each ratio must have the same simplest form.

All six numbers are integers, and only two are multiples of 5. So, neither of the quantities in the simplest form of each ratio is a multiple of 5. Therefore, the two choices which are multiples of 5 must be part of the same ratio so that the factors of 5 cancel when we simplify:

$30:140 = 3:14.$

The remaining numbers are 6, 9, 28, and 42. Of these, only 28 and 42 are multiples of 14, so the other two ratios are __:28 and __:42.

Filling these blanks with the two remaining numbers (6 and 9), we have $6:28 = 3:14$ and $9:42 = 3:14$.

So, the three equivalent ratios are

$30:140 = 6:28 = 9:42.$

We could write these ratios in any order within the equation, or flip the order of the quantities in each ratio. If you changed the order of the quantities in one ratio, be sure that you changed the order of the quantities in all the ratios!

$28:6 = 42:9 = 140:30.$

39. The only column that is completely filled tells us that the milk-to-butter ratio is $36:56$. This ratio simplifies to $9:14$, which allows us to fill in the left column:

ounces of milk	9	27	36	54	
ounces of butter	14		56		126

If we triple the amount of milk, then we must also triple the amount of butter to keep the same milk-to-butter ratio. Similarly, if we double the amount of milk, we must also double the amount of butter.

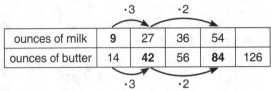

ounces of milk	9	27	36	54	
ounces of butter	14	**42**	56	**84**	126

Similarly, if we multiply the amount of butter by 9, we must also multiply the amount of milk by 9.

·9

ounces of milk	9	27	36	54	**81**
ounces of butter	14	**42**	56	**84**	126

·9

40. The original ratio of pigs to goats in the pen is $18:24 = 3:4$, so for every 3 pigs, there are 4 goats.

After the new animals enter the pen, there are $18+9 = 27$ pigs, which we can split into $27÷3 = 9$ groups of 3 pigs each. Then, each of the 9 groups also has 4 goats, so there are $9·4 = 36$ goats. Therefore, $36-24 = $ **12** goats entered the pen.

— *or* —

The ratio of pigs to goats is $18:24 = 3:4$. Since the entrance of the new animals did not change the pig-to-goat ratio, for every 3 pigs that entered the pen, 4 goats must have also entered.

Since $9 = 3·\underline{3}$ pigs entered the pen, $3·\underline{4} = $ **12** goats must have also entered the pen.

41. The ratio of gold to silver coins is $36:63 = 4:7$, so Kraken has 4 gold coins for every 7 silver coins.

After Kraken spends some of the coins, he has $36-8 = 28$ gold coins remaining, which we can split into $28÷4 = 7$ groups of 4 gold coins each. Then, each of the 7 groups also has 7 silver coins, so there are $7·7 = 49$ silver coins remaining. Therefore, Kraken spent $63-49 = $ **14** silver coins.

— *or* —

The ratio of gold to silver coins is $36:63 = 4:7$. Since spending some of the coins did not change the gold-to-silver ratio, for every 4 gold coins Kraken spent, he also spent 7 silver coins.

Since Kraken spent $8 = 2·\underline{4}$ gold coins, he spent $2·\underline{7} = $ **14** silver coins.

42. The original ratio of cups of cornstarch to cups of water is $\frac{3}{4} : \frac{1}{2} = 3:2$. Since Grogg wants to keep this cornstarch-to-water ratio, he must add 3 cups of cornstarch for every 2 cups of water he adds.

Since he adds 1 cup of cornstarch, and $1 = \frac{1}{3} \cdot 3$, we know that he must add $\frac{1}{3} \cdot 2 = \frac{2}{3}$ **cups** of water.

RATIOS & RATES — *Rectivide* — 46-47

For each of the following problems, there is only one solution.

43. The ratio of gray squares to white squares in the original rectangle is $3:3 = 1:1$.

We can split the rectangle into three smaller rectangles as shown below so that the ratio of gray squares to white squares is 1:1 in each smaller rectangle.

44. The ratio of gray squares to white squares in the original rectangle is $6:6 = 1:1$. So, we can make groups of 2 squares in each of the smaller rectangles, each with 1 gray square and 1 white square. Therefore, the area of each small rectangle is a multiple of 2.

We can split the rectangle into three smaller rectangles as shown below so that the ratio of gray squares to white squares is 1:1 in each smaller rectangle.

We use the same reasoning discussed in the previous problems to solve each Rectivide puzzle that follows.

45. $9:3 = 3:1$

46. $4:8 = 1:2 = 2:4$

47. $8:12 = 2:3 = 4:6$

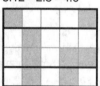

48. $15:3 = 5:1$

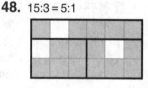

49. $6:12 = 1:2 = 2:4 = 3:6$

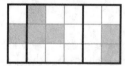

50. $9:9 = 2:2 = 3:3 = 4:4$

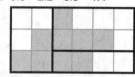

51. $9:9 = 1:1 = 2:2 = 6:6$

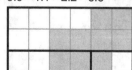

52. $16:4 = 4:1 = 8:2$

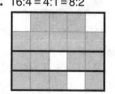

53. $5:15 = 1:3 = 3:9$

54. $10:10 = 1:1 = 4:4 = 5:5$

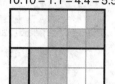

55. We can write $\frac{2}{3}$ with a denominator of 15 by multiplying the numerator and denominator by 5.

$$\frac{2}{3} \overset{\cdot 5}{\underset{\cdot 5}{=}} \frac{\mathbf{10}}{15}$$

56. We can write $\frac{5}{12}$ with a numerator of 15 by multiplying the numerator and denominator by 3.

$$\frac{15}{\mathbf{36}} \overset{\cdot 3}{\underset{\cdot 3}{=}} \frac{5}{12}$$

57. We can write $\frac{7}{8}$ with a denominator of 32 by multiplying the numerator and denominator by 4.

$$\frac{\mathbf{28}}{32} \overset{\cdot 4}{\underset{\cdot 4}{=}} \frac{7}{8}$$

58. We can write $\frac{19}{11}$ with a denominator of 88 by multiplying the numerator and denominator by 8.

$$\frac{19}{11} \overset{\cdot 8}{\underset{\cdot 8}{=}} \frac{\mathbf{152}}{88}$$

59. We can write $\frac{3}{7}$ with a denominator of 42 by multiplying the numerator and denominator by 6.

$$\frac{3}{7} \overset{\cdot 6}{\underset{\cdot 6}{=}} \frac{18}{42}$$

$\frac{3}{7} = \frac{18}{42}$, so $x = \mathbf{18}$.

— or —

To isolate the variable x, we multiply both sides of the equation by 42.

$$\frac{3}{7} \cdot 42 = \frac{x}{42} \cdot 42$$
$$\frac{3}{7} \cdot \overset{6}{\cancel{42}} = \frac{x}{\cancel{42}} \cdot \cancel{42}$$
$$\mathbf{18} = x.$$

60. We can write $\frac{13}{9}$ with a denominator of 45 by multiplying the numerator and denominator by 5.

$$\frac{13}{9} \overset{\cdot 5}{\underset{\cdot 5}{=}} \frac{\mathbf{65}}{45}$$

$\frac{13}{9} = \frac{65}{45}$, so $a = \mathbf{65}$.

— or —

To isolate the variable a, we multiply both sides of the equation by 45.

$$\frac{13}{9} \cdot 45 = \frac{a}{45} \cdot 45$$

$$\frac{13}{\cancel{9}} \cdot \cancel{45}^5 = \frac{a}{\cancel{45}} \cdot \cancel{45}$$

$$65 = a.$$

61. We simplify $\frac{21}{15} = \frac{7}{5}$. Then, we can write $\frac{7}{5}$ with a denominator of 35 by multiplying the numerator and denominator by 7.

$$\frac{49}{35} \overset{\cdot 7}{\underset{\cdot 7}{=}} \frac{7}{5}$$

$\frac{21}{15} = \frac{7}{5} = \frac{49}{35}$, so $w = \textbf{49}$.

— *or* —

We simplify $\frac{21}{15}$ to $\frac{7}{5}$. Then, to isolate the variable w, we multiply both sides of the equation by 35.

$$\frac{w}{35} \cdot 35 = \frac{7}{5} \cdot 35$$

$$\frac{w}{\cancel{35}} \cdot \cancel{35} = \frac{7}{\cancel{5}} \cdot \cancel{35}^7$$

$$w = \textbf{49}.$$

62. We cannot multiply the numerator and denominator of $\frac{9}{4}$ by an integer to get an equivalent fraction with denominator 21. So, we isolate the variable.

To isolate the variable we multiply both sides of the equation by 21.

$$\frac{m}{21} \cdot \cancel{21} = \frac{9}{4} \cdot 21$$

$$m = \frac{9 \cdot 21}{4}$$

$$m = \frac{189}{4} = \textbf{47}\frac{1}{4}.$$

63. We can eliminate the denominators of $\frac{2}{3}$ and $\frac{15}{m}$ by multiplying both sides of the equation by a common multiple of their denominators: $3m$.

$$\frac{2}{3} \cdot \cancel{3}m = \frac{15}{\cancel{m}} \cdot 3\cancel{m}$$

$$2 \cdot m = 15 \cdot 3$$

This gives $2m = 45$.

$$2m = 45$$

We divide both sides by 2 to get $m = \frac{45}{2} = 22\frac{1}{2}$.

$$m = \frac{45}{2}$$

— *or* —

For any equation $\frac{a}{b} = \frac{c}{d}$, we have $ad = bc$.

So, $\frac{2}{3} = \frac{15}{m}$ gives $2 \cdot m = 15 \cdot 3$.

We simplify, then solve for m:

$$2 \cdot m = 15 \cdot 3$$

$$2m = 45$$

$$m = \frac{45}{2} = 22\frac{1}{2}.$$

64. We solve for s as shown below.

$$\frac{12}{s} = \frac{5}{8}$$

$$\frac{12}{\cancel{s}} \cdot 8\cancel{s} = \frac{5}{\cancel{8}} \cdot \cancel{8}s$$

$$12 \cdot 8 = 5 \cdot s$$

$$96 = 5s$$

$$\frac{96}{5} = s.$$

So, $s = \frac{96}{5} = \textbf{19}\frac{1}{5}$.

65. We solve for c as shown below.

$$\frac{7}{4} = \frac{15}{c}$$

$$\frac{7}{4} \cdot 4c = \frac{15}{\cancel{c}} \cdot 4\cancel{c}$$

$$7 \cdot c = 15 \cdot 4$$

$$7c = 60$$

$$c = \frac{60}{7} = \textbf{8}\frac{4}{7}.$$

66. We solve for a as shown below.

$$\frac{14}{a} = \frac{4}{9}$$

$$\frac{14}{\cancel{a}} \cdot 9\cancel{a} = \frac{4}{\cancel{9}} \cdot \cancel{9}a$$

$$14 \cdot 9 = 4 \cdot a$$

$$126 = 4a$$

$$\frac{126}{4} = a.$$

So, $a = \frac{126}{4} = \frac{63}{2} = \textbf{31}\frac{1}{2}$.

67. We solve for z as shown below.

$$\frac{2}{5} = \frac{15}{z}$$

$$\frac{2}{5} \cdot 5z = \frac{15}{\cancel{z}} \cdot 5\cancel{z}$$

$$2 \cdot z = 15 \cdot 5$$

$$2z = 75$$

$$z = \frac{75}{2} = \textbf{37}\frac{1}{2}.$$

68. We solve for v as shown below.

$$\frac{3}{10} = \frac{10}{v}$$

$$\frac{3}{10} \cdot 10v = \frac{10}{\cancel{v}} \cdot 10\cancel{v}$$

$$3 \cdot v = 10 \cdot 10$$

$$3v = 100$$

$$v = \frac{100}{3} = \textbf{33}\frac{1}{3}.$$

69. We solve for n as shown below.

$$\frac{8}{n} = \frac{11}{6}$$

$$\frac{8}{\cancel{n}} \cdot 6\cancel{n} = \frac{11}{\cancel{6}} \cdot \cancel{6}n$$

$$8 \cdot 6 = 11 \cdot n$$

$$48 = 11n$$

$$\frac{48}{11} = n.$$

So, $n = \frac{48}{11} = \textbf{4}\frac{4}{11}$.

70. We solve for r as shown below.

$$\frac{10}{7} = \frac{6}{r}$$

$$\frac{10}{7} \cdot 7r = \frac{6}{\cancel{r}} \cdot 7\cancel{r}$$

$$10 \cdot r = 6 \cdot 7$$

$$10r = 42$$

$$r = \frac{42}{10} = \frac{21}{5} = \textbf{4}\frac{1}{5}.$$

In these problems, you may have flipped the order of the quantities in the ratios to arrive at the same final answer.

71. Suppose there are v vowels in the sentence. Then, the ratio of 40 consonants to v vowels is 40:v. Since there are 5 consonants for every 4 vowels, we have 40:v = 5:4.

Since 40:32 = 5:4, we have v = 32. So, the sentence has **32** vowels.

— *or* —

We express 40:v = 5:4 with fractions.

$$\frac{40}{v} = \frac{5}{4} \quad \begin{matrix} \leftarrow \text{ consonants} \\ \leftarrow \text{ vowels} \end{matrix}$$

We multiply both sides by 4v to solve for v as shown.

$$\frac{40}{v} \cdot 4v = \frac{5}{4} \cdot 4v$$
$$40 \cdot 4 = 5 \cdot v$$
$$160 = 5v$$
$$\frac{160}{5} = v$$

So, the sentence has $\frac{160}{5}$ = **32** vowels.

72. Suppose Cori needs c cups of cocoa. Then, the ratio of 5 cups of sugar to c cups of cocoa is 5:c. Since we want 9 parts sugar for every 2 parts cocoa, we have 5:c = 9:2, which gives

$$\frac{5}{c} = \frac{9}{2} \quad \begin{matrix} \leftarrow \text{ sugar} \\ \leftarrow \text{ cocoa} \end{matrix}$$

We multiply both sides by 2c to solve for c as shown.

$$\frac{5}{c} \cdot 2c = \frac{9}{2} \cdot 2c$$
$$5 \cdot 2 = 9 \cdot c$$
$$10 = 9c$$
$$\frac{10}{9} = c$$

So, Cori needs $\frac{10}{9} = 1\frac{1}{9}$ cups of cocoa.

73. Suppose the Blorgbeast's hoof is w inches wide. Then, the ratio of w inches of hoof width to 5 inches of hoof height is w:5. Since the ratio of hoof width to hoof height is 3:8, we have w:5 = 3:8, which gives

$$\frac{w}{5} = \frac{3}{8} \quad \begin{matrix} \leftarrow \text{ hoof width} \\ \leftarrow \text{ height} \end{matrix}$$

We multiply both sides by 5 to solve for w as shown.

$$\frac{w}{5} \cdot 5 = \frac{3}{8} \cdot 5$$
$$w = \frac{15}{8}$$

So, the hoof is $\frac{15}{8} = 1\frac{7}{8}$ inches wide.

74. Suppose we need to add s ounces of sugar. Then, the ratio of s ounces of sugar to 16 ounces of water is s:16. Since we want 2 parts sugar for every 7 parts water, we have s:16 = 2:7, which gives

$$\frac{s}{16} = \frac{2}{7} \quad \begin{matrix} \leftarrow \text{ sugar} \\ \leftarrow \text{ water} \end{matrix}$$

Then, we multiply both sides by 16 to solve for s as shown.

$$\frac{s}{16} \cdot 16 = \frac{2}{7} \cdot 16$$
$$s = \frac{32}{7}$$

So, $\frac{32}{7} = 4\frac{4}{7}$ ounces of sugar should be added to 16 ounces of water to make hummingbeast nectar.

75. Two sevenths of Terry's total hits are home runs. Since *two* sevenths of his total hits is 28 hits, *one* seventh of his total hits is 14 hits.

Therefore, Terry hit the ball 7·14 = 98 times, and he did not hit a home run 98−28 = **70** times.

— *or* —

Since Terry hits a home run 2 out of every 7 times that he hits the baseball, his ratio of home-run hits to non-home-run hits is 2:5.

Suppose Terry made h non-home-run hits. Then, the ratio of 28 home-run hits to h non-home-run hits is 28:h. Since he made 2 home-run hits for every 5 non-home-run hits, we have 2:5 = 28:h, which gives

$$\frac{2}{5} = \frac{28}{h} \quad \begin{matrix} \leftarrow \text{ home-run hits} \\ \leftarrow \text{ non-home-run hits} \end{matrix}$$

Since $\frac{2}{5} = \frac{28}{70}$, we have h = 70. Therefore, Terry had **70** non-home-run hits.

76. The area of triangle ABC is $\frac{6 \cdot 6}{2}$ = 18 square inches.

The ratio of the area of triangle ABC to the area of triangle DEF is 3:5. Since the area of triangle ABC is 18 = 6·3 square inches, the area of triangle DEF is 6·5 = 30 square inches.

The area of a right triangle is half the product of its leg lengths, so the product of the lengths of EF and ED is 30·2 = 60. Since EF is 6 inches long, ED is 60÷6 = **10** inches long.

— *or* —

The area of a triangle is equal to half the product of its base and height. So, the ratio of the areas of two triangles that have the same base is the same as the ratio of their heights.

We consider AB and EF as the bases of their triangles, so BC and DE are the heights. Since the ratio of the area of triangle ABC to the area of triangle DEF is 3:5, and their bases are the same length, the ratio of the height of triangle ABC to the height of triangle DEF 3:5.

Since BC is 6 = 2·3 inches long, DE is 2·5 = **10** inches long.

77. Since the ratio of the weight of a biffo to the weight of a triffo is 2:3, the ratio of the weight of a triffo to the weight of a biffo is 3:2. This means that the weight of a triffo divided by the weight of a biffo is $\frac{3}{2}$.

So, the weight of a triffo is $\frac{3}{2}$ the weight of a biffo.

Since 30 biffos weigh 40 pounds, 30 triffos weigh $\frac{3}{2}$ times as much. Therefore, 30 triffos weigh $\frac{3}{2} \cdot 40$ = **60** pounds.

— *or* —

Since the ratio of the weight of one biffo to the weight of one triffo is 2:3, the ratio of the weight of 30 biffos to the weight of 30 triffos is also 2:3.

Suppose 30 triffos weigh w pounds. Since 30 biffos weigh 40 pounds, the ratio of biffo weight to triffo weight is 2:3 = 40:w, which gives

$$\frac{2}{3} = \frac{40}{w} \quad \leftarrow \text{biffo weight} \\ \quad\quad\quad \leftarrow \text{triffo weight}$$

Since $\frac{2}{3} = \frac{40}{60}$, we have $w = 60$. So, 30 triffos weigh **60** pounds.

78. Since $a{:}b = 3{:}5$, we know that a is $\frac{3}{8}$ of $a+b$, and b is $\frac{5}{8}$ of $a+b$.

Since $a+b = 100$, we have
$$a = \frac{3}{8} \cdot 100 = \frac{300}{8} = \frac{75}{2} = \mathbf{37\frac{1}{2}}.$$

Similarly, $b = \frac{5}{8} \cdot 100 = \frac{500}{8} = \frac{125}{2} = \mathbf{62\frac{1}{2}}.$

— *or* —

The ratio of a to b is 3:5. So, for some value of x, we have $a = 3x$ and $b = 5x$.

Since $a+b = 100$, we have
$$3x + 5x = 100$$
$$8x = 100$$
$$x = \frac{100}{8}$$
$$x = \frac{25}{2}.$$

So, we have
$$a = 3x = 3 \cdot \frac{25}{2} = \frac{75}{2} = \mathbf{37\frac{1}{2}}, \text{ and}$$
$$b = 5x = 5 \cdot \frac{25}{2} = \frac{125}{2} = \mathbf{62\frac{1}{2}}.$$

Check: $37\frac{1}{2}{:}62\frac{1}{2} = 3{:}5$, and $37\frac{1}{2} + 62\frac{1}{2} = 100.$ ✓

79. Of the goals scored by Ben and Alfie, $\frac{5}{9}$ were scored by Ben and $\frac{4}{9}$ were scored by Alfie.

So, Ben outscored Alfie by $\frac{5}{9} - \frac{4}{9} = \frac{1}{9}$ of their total goals. Since Ben scored 16 more goals than Alfie, 16 is one ninth of the total goals scored by Ben and Alfie.

Together, they scored $9 \cdot 16 = 144$ goals. Ben scored $\frac{5}{9} \cdot 144 = 80$, and Alfie scored $\frac{4}{9} \cdot 144 = \mathbf{64}$. Ben scored $80 - 64 = 16$ more goals. ✓

— *or* —

For every 5 goals Ben scored, Alfie scored 4. So, for some value of x, Ben scored $5x$ goals and Alfie scored $4x$ goals. Therefore, Ben outscored Alfie by $5x - 4x = x$ goals. Since Ben scored 16 more goals than Alfie, we have $x = 16$. So, Ben scored $5x = 5 \cdot 14 = 80$ goals, and Alfie scored $4x = 4 \cdot 16 = \mathbf{64}$.

Check: The ratio of Ben's goals to Alfie's goals is 80:64 = 5:4, and Ben scored $80 - 64 = 16$ more goals. ✓

RATIOS & RATES
Geometry   52–53

80. The ratio of the short side of the first rectangle to the short side of the second rectangle is 5:15.

The ratio of the long side of the first rectangle to the long side of the second rectangle is 8:w.

The rectangles are similar, so 5:15 = 8:w, which gives $\frac{5}{15} = \frac{8}{w}$. Solving for w, we have $w = \mathbf{24}$.

81. The ratio of the short side of the first rectangle to the short side of the second rectangle is 14:8.

The ratio of the long side of the first rectangle to the long side of the second rectangle is 21:r.

The rectangles are similar, so 14:8 = 21:r, which gives $\frac{14}{8} = \frac{21}{r}$. Solving for r, we have $r = \mathbf{12}$.

82. The ratio of the short leg of the first triangle to the short leg of the second triangle is t:6.

The ratio of the long leg of the first triangle to the long leg of the second triangle is 10:18.

The triangles are similar, so t:6 = 10:18, which gives $\frac{t}{6} = \frac{10}{18}$. Solving for t, we have $t = \frac{30}{9} = \frac{10}{3} = \mathbf{3\frac{1}{3}}$.

83. The ratio of the shortest side of the first triangle to the shortest side of the second triangle is 6:9.

The ratio of the longest side (the side opposite the obtuse angle) of the first triangle to the longest side of the second triangle is 14:x.

The triangles are similar, so 6:9 = 14:x, which gives $\frac{6}{9} = \frac{14}{x}$. Solving for x, we have $x = \mathbf{21}$.

The ratio of the remaining side of the first triangle to the remaining side of the second triangle is 10:y.

The triangles are similar, so 6:9 = 10:y, which gives $\frac{6}{9} = \frac{10}{y}$. Solving for y, we have $y = \mathbf{15}$.

84. We can divide the square faces of the small cube into $5 \cdot 5 = 25$ little squares, as shown on the left below. Since the ratio of the edge length of the small cube to the edge length of the large cube is 5:7, we can place 7 of the same little squares along each side of a face of the large cube. So, we can divide each face of the large cube into $7 \cdot 7 = 49$ little squares.

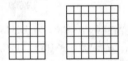

Each cube has 6 congruent faces. So, the surface area of the small cube is $6 \cdot 25 = 150$ of these little squares, and the surface area of the large cube is $6 \cdot 49 = 294$ of these little squares. Therefore, the ratio of the surface area of the small cube to the surface area of the large cube is 150:294 = **25:49**.

— *or* —

Since each cube has 6 congruent faces, the ratio of the surface area of the small cube to the surface area of the large cube is equal to the ratio of the area of one face of the smaller cube to the area of one face of the large cube.

Above, we found that when we split the faces of these cubes into little squares of the same size, a face of the small cube has $5 \cdot 5 = 25$ little squares, and a face of the large cube has $7 \cdot 7 = 49$ little squares.

So, the ratio of the surface area of the small cube to the surface area of the large cube is **25:49**.

— *or* —

For every 5 units in the edge length of the small cube, there are 7 units in the edge length of the large cube.

So, for some number of units x, the small cube has $5x$-unit edges, and the large cube has $7x$-unit edges.

Since each cube has 6 congruent faces, the ratio of the surface area of the small cube to the surface area of the large cube is equal to the ratio of the area of one face of the small cube to the area of one face of the large cube.

The area of a face of a cube with edge length $5x$ units is $5x \cdot 5x = 25x^2$ square units. The area of a face of a cube with edge length $7x$ units is $7x \cdot 7x = 49x^2$ square units.

So, the ratio of the surface area of the small cube to the surface area of the large cube is $25x^2 : 49x^2$. We divide both quantities by x^2 to simplify this ratio to **25:49**.

85. **a.** The ratio of model inches to actual feet is
$$3\tfrac{3}{4} : 50 = \tfrac{15}{4} : 50 = 15:200 = 3:40.$$

Suppose the length of the original train car is f feet. Since 3 model inches represent 40 actual feet, and 6 model inches represent f actual feet, we have $3:40 = 6:f$, which gives
$$\frac{3}{40} = \frac{6}{f} \quad \leftarrow \text{ model inches} \\ \qquad\qquad \leftarrow \text{ actual feet}$$
Since $\frac{3}{40} = \frac{6}{80}$, we have $f = 80$. The actual train car is **80 feet long**.

b. In part (a), we found that the ratio of model inches to actual feet is 3:40.

Suppose the model track is m inches long. Since 3 model inches represent 40 actual feet, and m model inches represent 12,000 feet, we have $3:40 = m:12,000$, which gives
$$\frac{3}{40} = \frac{m}{12,000} \quad \leftarrow \text{ model inches} \\ \qquad\qquad\qquad \leftarrow \text{ actual feet}$$

Multiplying both sides by 12,000, we have $m = \frac{3}{40} \cdot 12,000 = 3 \cdot 300 = 900$. So, Chis's model is 900 inches long.

There are 12 inches in a foot, so Chris's model track is $900 \div 12 = $ **75 feet long**.

86. **a.** Suppose Griffinburg and North Pegasus are actually m miles apart. Since 3 map inches represent 25 actual miles, and $2\tfrac{1}{4}$ map inches represent m miles, we have $3:25 = 2\tfrac{1}{4} : m$.

We eliminate the fraction in the ratio on the right side of the equation by multiplying both quantities by 4:
$$2\tfrac{1}{4} : m = 9 : 4m.$$

Therefore, $3:25 = 9:4m$, which gives
$$\frac{3}{25} = \frac{9}{4m} \quad \leftarrow \text{ map inches} \\ \qquad\qquad \leftarrow \text{ actual miles}$$

Since $\frac{3}{25} = \frac{9}{75}$, we have $4m = 75$. So, $m = \frac{75}{4}$, and the two cities are $\frac{75}{4} = \mathbf{18\tfrac{3}{4}}$ miles apart.

— *or* —

Since 3 map inches represent 25 actual miles, every $\frac{3}{3} = 1$ map-inch represents $\frac{25}{3} = 8\tfrac{1}{3}$ miles. Since Griffinburg and North Pegasus are $2\tfrac{1}{4}$ map-inches apart, these two cities are actually $2\tfrac{1}{4} \cdot 8\tfrac{1}{3} = \frac{9}{4} \cdot \frac{25}{3} = \frac{75}{4} = \mathbf{18\tfrac{3}{4}}$ miles apart.

b. The ratio of map inches to actual miles is 3:25.

Suppose Unicorpia is m map inches away from West Mermaid. Since 3 map inches represent 25 miles, and m map inches represent 40 miles, we have $3:25 = m:40$, which gives
$$\frac{3}{25} = \frac{m}{40} \quad \leftarrow \text{ map inches} \\ \qquad\qquad \leftarrow \text{ actual miles}$$

Then, we multiply both sides by 40 to solve for m as shown.
$$\frac{3}{25} \cdot 40 = \frac{m}{40} \cdot 40 \\ \frac{3 \cdot 40}{25} = m \\ \frac{24}{5} = m$$

So, West Mermaid would be $\frac{24}{5} = \mathbf{4\tfrac{4}{5}}$ inches from Unicorpia on the map.

RATIOS & RATES
Multi-Part Ratios   54-55

87. The ratio of red to blue to yellow shirts is 18:30:45. We simplify the ratio by dividing 18, 30, and 45 by their greatest common factor, 3. So, $18:30:45 = $ **6:10:15**.

88. There are $60 - 15 - 14 = 31$ fifth graders in the parade. The greatest common factor of 15, 14, and 31 is 1, so the ratio of third to fourth to fifth graders is **15:14:31**.

89. From the given ratio, we know that we can make groups of $8+5+3 = 16$ fish, each with 5 goldfish. So, $\frac{5}{16}$ of the fish are goldfish.

90. From the Fuji-to-Gala ratio, we know that $\frac{3}{8}$ of Rolfe's apple trees are Fuji and $\frac{5}{8}$ are Gala.

From the apple-to-peach ratio, we know that $\frac{4}{7}$ of his trees are apple trees, and $\frac{3}{7}$ are peach trees.

Therefore, $\frac{3}{8} \cdot \frac{4}{7} = \frac{3}{14}$ of all Rolfe's trees are Fuji apple trees, $\frac{5}{8} \cdot \frac{4}{7} = \frac{5}{14}$ of his trees are Gala apple trees, and $\frac{3}{7} = \frac{6}{14}$ of his trees are peach trees.

So, we can make groups of 14 trees, each with 3 Fuji, 5 Gala, and 6 peach. The ratio of Fuji to Gala to peach trees is **3:5:6**.

— *or* —

The ratio of Fuji trees to Gala trees is 3:5. So, we can make groups of 8 apple trees, each with 3 Fuji and 5 Gala.

The ratio of apple trees to peach trees is 4:3. Since apple trees are easier to think about in groups of 8 than in groups of 4, we write the apple-to-peach ratio as $4:3 = 8:6$.

So, we can make groups of $8+6=14$ trees, each with 8 apple trees and 6 peach trees. Three of the apple trees are Fuji, while the other 5 are Gala.

Apple		Peach
8		6
Fuji	Gala	Peach
3	5	6

The ratio of Fuji to Gala to peach trees is **3:5:6**.

91. The ratio of butter to sugar to flour is
$$1\tfrac{1}{2}:2\tfrac{1}{2}:2\tfrac{2}{3}=\tfrac{3}{2}:\tfrac{5}{2}:\tfrac{8}{3}.$$
To write this as a ratio of integers, we multiply every quantity by the least common multiple of the denominators, 6. So, $\tfrac{3}{2}:\tfrac{5}{2}:\tfrac{8}{3}=$ **9:15:16**.

92. From the given ratio, we know that for every 5 units that the prism is long, it is 6 units wide and 8 units tall.

Since the prism is $10\cdot\underline{6}=60$ cm wide, it is $10\cdot\underline{5}=50$ cm long and $10\cdot\underline{8}=80$ cm tall.

A 50-by-60-by-80-cm prism has two 50-by-60-cm faces, two 60-by-80-cm faces, and two 50-by-80-cm faces. Therefore, the surface area of the prism is
$$2\cdot(50\cdot60)+2\cdot(60\cdot80)+2\cdot(50\cdot80)$$
$$=2\cdot(3,000+4,800+4,000)$$
$$=2\cdot11,800$$
$$=\textbf{23,600 square centimeters.}$$

93. From the given ratio, we know that $\tfrac{2}{6}=\tfrac{1}{3}$ of the marbles in the bag are blue. Since the 16 blue marbles are one third of the marbles in the bag, the bag holds a total of $16\cdot3=$ **48** marbles.

94. From the green-to-yellow frog ratio, we know that $\tfrac{3}{5}$ of the frogs are green and $\tfrac{2}{5}$ are yellow.

From the frog-to-toad ratio, we know that $\tfrac{3}{8}$ of these animals are frogs and $\tfrac{5}{8}$ are toads.

Therefore, $\tfrac{3}{5}\cdot\tfrac{3}{8}=\tfrac{9}{40}$ of the animals are green frogs, $\tfrac{2}{5}\cdot\tfrac{3}{8}=\tfrac{3}{20}=\tfrac{6}{40}$ of the animals are yellow frogs, and $\tfrac{5}{8}=\tfrac{25}{40}$ of the animals are toads.

So, we can make groups of 40 animals, each with 9 green frogs, 6 yellow frogs, and 25 toads. The ratio of green frogs to yellow frogs to toads is **9:6:25**.

— or —

The ratio of green frogs to yellow frogs is 3:2. So, we can make groups of 5 frogs, each with 3 green and 2 yellow.

The ratio of frogs to toads is 3:5. Since frogs are easier to think about in groups of 5 than in groups of 3, we write the frog-to-toad ratio as $3:5=15:25$.

So, we can make groups of $15+25=40$ animals, each with 15 frogs and 25 toads. In each group of 15 frogs, $3\cdot\underline{3}=9$ are green and $3\cdot\underline{2}=6$ are yellow.

Frogs		Toads
15		25
Green	Yellow	Toads
9	6	25

The ratio of green frogs to yellow frogs to toads is **9:6:25**.

95. From the given ratio, we know that a is $\tfrac{2}{14}=\tfrac{1}{7}$ of $a+b+c$, b is $\tfrac{9}{14}$ of $a+b+c$, and c is $\tfrac{3}{14}$ of $a+b+c$.

Since $a+b+c=70$, we have $a=\tfrac{1}{7}\cdot70=10$.

Similarly, $b=\tfrac{9}{14}\cdot70=45$ and $c=\tfrac{3}{14}\cdot70=15$.

Therefore, $abc=10\cdot45\cdot15=\textbf{6,750}$.

— or —

The ratio of a to b to c is 2:9:3. So, for some value of x, we have $a=2x$, $b=9x$, and $c=3x$.

Since $a+b+c=70$, we have
$$2x+9x+3x=70$$
$$14x=70$$
$$x=\frac{70}{14}$$
$$x=5.$$

So, we have
$$a=2x=2\cdot5=10,$$
$$b=9x=9\cdot5=45,\text{ and}$$
$$c=3x=3\cdot5=15.$$

Therefore, $abc=10\cdot45\cdot15=\textbf{6,750}$.

Check: $10:45:15=2:9:3$, and $10+45+15=70$. ✓

96. From the given ratio, we know that for every 2 inches that the prism is long, it is 5 inches wide and 8 inches tall.

So, for some number of inches x, the prism is $2x$ inches long, $5x$ inches wide, and $8x$ inches tall, and its volume is
$$2x\cdot5x\cdot8x=(2\cdot5\cdot8)\cdot(x\cdot x\cdot x)=80x^3\text{ cubic inches.}$$

The volume of the prism is 2,160 cubic inches, so we have $80x^3=2,160$. Dividing both sides of this equation by 80 gives $x^3=27$. Since $3^3=27$, we have $x=3$.

So, the prism is $2x=2\cdot3=$ **6 inches long**, $5x=5\cdot3=$ **15 inches wide**, and $8x=8\cdot3=$ **24 inches tall**.

Check: The volume of a 6-by-15-by-24-inch rectangular prism is $6\cdot15\cdot24=2,160$ cubic inches. ✓

Ratiotile   56-57

97. The ratio of the triangles that touch the left square is $6:15=2:5$, and the ratio of the triangles that touch the right square is $6:16=3:8$.

98. Every triangle contains an integer, so the ratio in the left square tells us that the smaller number in the triangles that touch it is a multiple of 2.

Since 9 is not a multiple of 2, it must be the larger number in the triangles that touch this square.

$\boxed{6}:9=2:3$, so the empty triangle contains 6.

Then, the ratio of the triangles that touch the right square is $6:8=3:4$.

99. The ratio of the triangles that touch the left square is 2:9. Since 4 is a multiple of 2 but not 9, it must be the smaller number in the triangles that touch this square.

4:⬚18⬚ = 2:9, so the empty triangle contains 18.

Then, the ratio of the triangles that touch the right square is 6:18 = 1:3.

100. The ratio of the triangles that touch the square on the left is 1:2.

If 2 is the larger number, we have ⬚1⬚:2 = 1:2. If 2 is the smaller number, then we have 2:⬚4⬚ = 1:2. So, the center triangle is 1 or 4.

The ratio of the triangles that touch the right square is 2:3. There is no integer that has a 2:3 ratio with 1. So, the center triangle is 4.

There is no integer smaller than 4 that has a 2:3 ratio with 4. However, if 4 is the smaller number in the 2:3 ratio, then we have 4:⬚6⬚ = 2:3. So, the upper-right triangle is 6.

We use the strategies discussed in the previous solutions to solve the Ratiotile puzzles that follow.

101.

102.

103.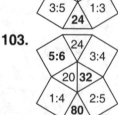

104. The ratio of the triangles that touch the top-right square 2:3.

If 12 is the larger number, we have ⬚8⬚:12 = 2:3. If 12 is the smaller number, we have 12:⬚18⬚ = 2:3. So, the top-center triangle is 8 or 18.

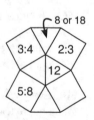

Suppose we place 8 in the top-center triangle. The ratio of the triangles that touch the top-left square is 3:4. Since 8 is a multiple of 4 but not 3, it is the larger number, and we have ⬚6⬚:8 = 3:4.

However, the 6 triangle also touches the bottom-left square. We cannot make a ratio with 6 and another integer that simplifies to 5:8. ✗

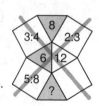

So, the top-center triangle is 18.

The ratio of the triangles that touch the top-left square is 3:4. Since 18 is a multiple of 3 but not 4, it is the smaller number, and we have 18:⬚24⬚ = 3:4.

The ratio of the triangles around the bottom-left square is 5:8. Since 24 is a multiple of 8 but not 5, it is the larger number, and we have ⬚15⬚:24 = 5:8.

Finally, the ratio of the triangles that touch the bottom-right square is 12:15 = 4:5.

105. The ratio of the triangles that touch the top-left square is 1:3. If 12 is the larger number, we have ⬚4⬚:12 = 1:3. If 12 is the smaller number, we have 12:⬚36⬚ = 1:3. So, the top triangle is 4 or 36.

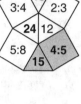

The ratio in the top-right square tells us that the smaller number in the triangles that touch it is a multiple of 7.

Neither 4 nor 36 is a multiple of 7, but the other triangle that touches this square contains 21, which is a multiple of 7. So, 21 is the smaller number in the triangles that touch the top-right square, and 36 is the larger number.

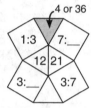

The ratio of the triangles that touch the top-right square is 21:36 = 7:⬚12⬚.

The ratio of the triangles that touch the bottom-right square is 3:7. If 21 is the larger number, we have ⬚9⬚:21 = 3:7. If 21 is the smaller number, we have 21:⬚49⬚ = 3:7. So, the bottom triangle is 9 or 49.

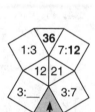

However, if 12 and 49 are the two triangles that touch the bottom-left square, their ratio cannot simplify to 3:⬚, with an integer in the blank.

Therefore, the bottom triangle is 9, and the ratio of the triangles that touch the bottom-left square is 9:12 = 3:⬚4⬚.

106. The ratio in the left square is either 1:5, 2:5, 3:5, or 4:5. Since 16 is not a multiple of 5, we know 16 is the smaller number in the ratio. Also, since 16 is not a multiple of 3, the ratio of the triangles that touch the left square cannot be 3:5.

We consider the remaining possibilities.

<u>1:5</u>: Since 16:⬚80⬚ = 1:5, the bottom triangle can be 80.

<u>2:5</u>: Since 16:⬚40⬚ = 2:5, the bottom triangle can be 40.

<u>4:5</u>: Since 16:⬚20⬚ = 4:5, the bottom triangle can be 20.

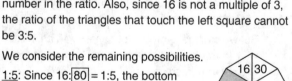

[20]:30 = 2:3, but 30:[40] = 3:4 and 30:[80] = 3:8. Only 20 can fill the bottom triangle so that the ratio in the right square simplifies to ☐:3.

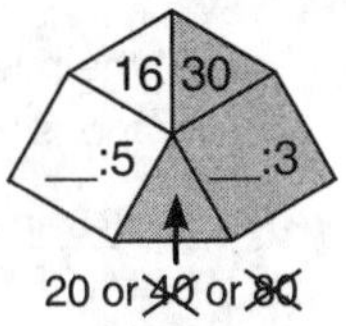
16
30
__:5
__:3
20 or 40 or 80

So, the center triangle is 20. Then, the ratio of the triangles that touch the left square is 16:20 = 4:5, and the ratio of the triangles that touch the right square is 20:30 = 2:3.

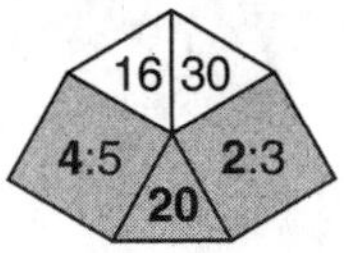
16
30
4:5
2:3
20

We use the strategies discussed in the previous solutions to solve the Ratiotile puzzles that follow.

107.

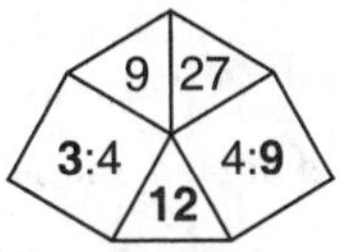
9
27
3:4
4:9
12

108.

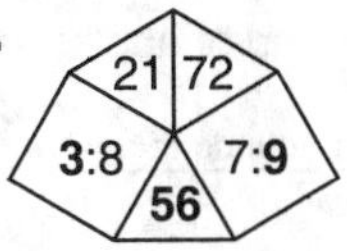
21
72
3:8
7:9
56

109. The ratio of the triangles that touch the top-left square is 3:5:8. Since 20 is not a multiple of 3 or 8, we know 20 is the middle value of this ratio. So, we have [12]:20:[32] = 3:5:8, and the empty triangles that touch the top-left square are 12 and 32.

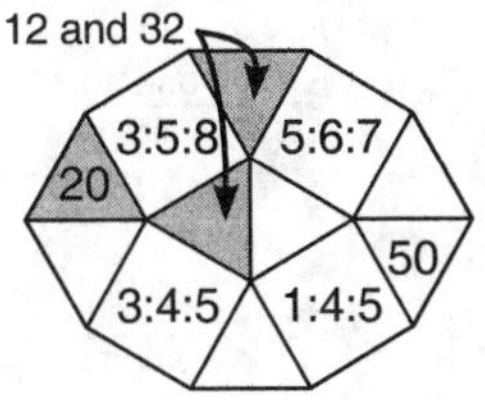
12 and 32
3:5:8
5:6:7
20
50
3:4:5
1:4:5

The ratio of the triangles that touch the top-right square is 5:6:7. Since 32 is not a multiple of 5, 6, or 7, we cannot place 32 in the top triangle.

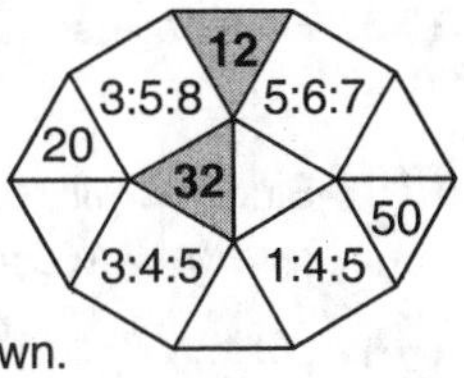
12
3:5:8
5:6:7
20
32
50
3:4:5
1:4:5

So, we place the 12 and 32 as shown.

The ratio of the triangles that touch the top-right square is 5:6:7. Since 12 is not a multiple of 5 or 7, it is the middle value of this ratio. Then, we have [10]:12:[14] = 5:6:7, and the empty triangles that touch the top-right square are 10 and 14.

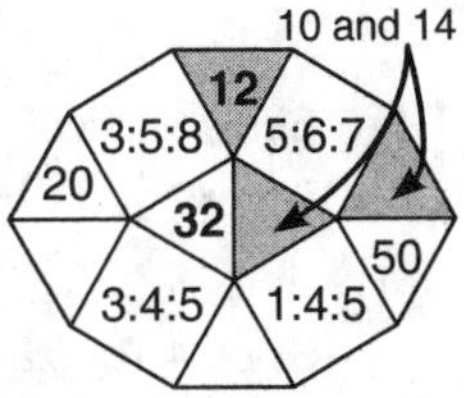
10 and 14
12
3:5:8
5:6:7
20
32
50
3:4:5
1:4:5

The shaded empty triangle to the right also touches the bottom-right square. There is no ratio of three integers with 14 and 50 that simplifies to 1:4:5.

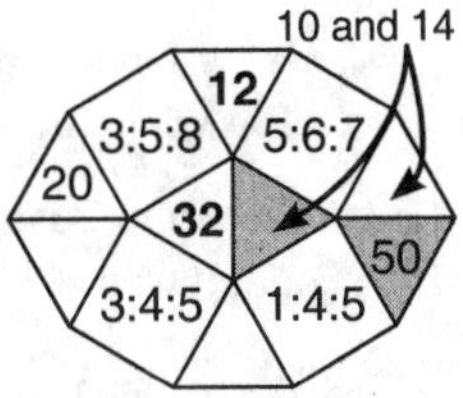
10 and 14
12
3:5:8
5:6:7
20
32
50
3:4:5
1:4:5

Therefore, we place the 10 and 14 as shown.

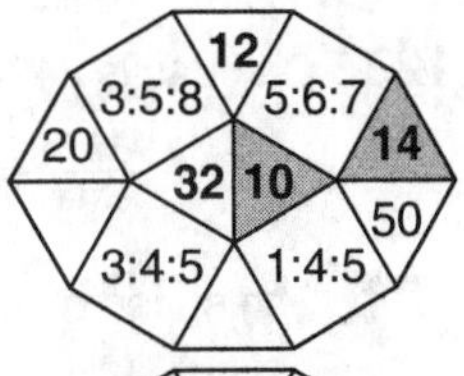
12
3:5:8
5:6:7
20
14
32
10
50
3:4:5
1:4:5

10:[40]:50 is the only ratio of three integers with 10 and 50 that simplifies to 1:4:5. So, the bottom triangle is 40.

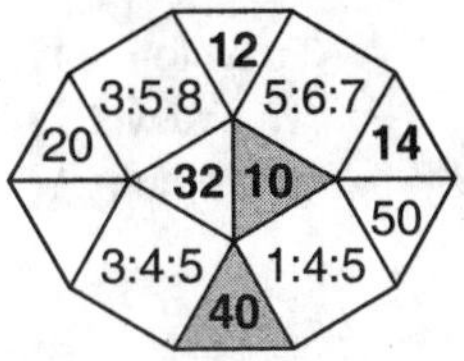
12
3:5:8
5:6:7
20
14
32
10
50
3:4:5
1:4:5
40

Finally, [24]:32:40 is the only ratio of three integers with 32 and 40 that simplifies to 3:4:5. So, the bottom-left triangle is 24.

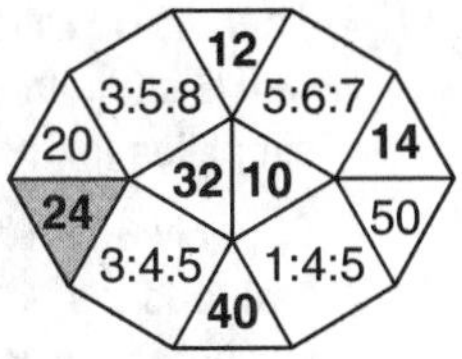
12
3:5:8
5:6:7
20
14
32
10
24
50
3:4:5
1:4:5
40

110. The ratio of the triangles that touch the bottom-left square is 2:3:5. Since 9 is not a multiple of 2 or 5, the middle value of this ratio is 9. Then, we have [6]:9:[15] = 2:3:5, and the empty triangles that touch the bottom-left square are 6 and 15.

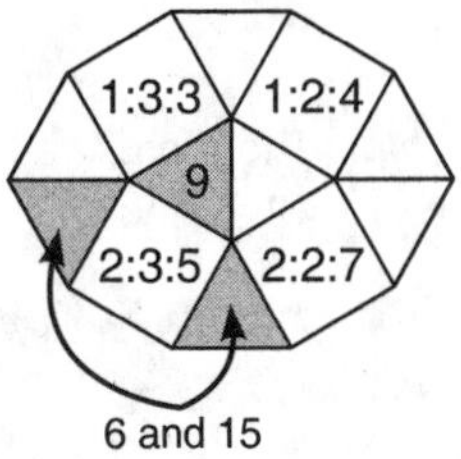
1:3:3
1:2:4
9
2:3:5
2:2:7
6 and 15

The ratio of the triangles that touch the bottom-right square is 2:2:7. Since 15 is not a multiple of 2 or 7, we cannot place 15 in the bottom triangle.

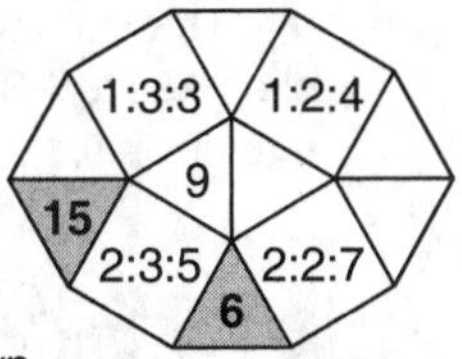
1:3:3
1:2:4
9
15
2:3:5
2:2:7
6

So, we place the 6 and 15 as shown.

The ratio of the triangles that touch the bottom-right square is 2:2:7. Since 6:[6]:[21] = 2:2:7, the two other triangles that touch this square are 6 and 21.

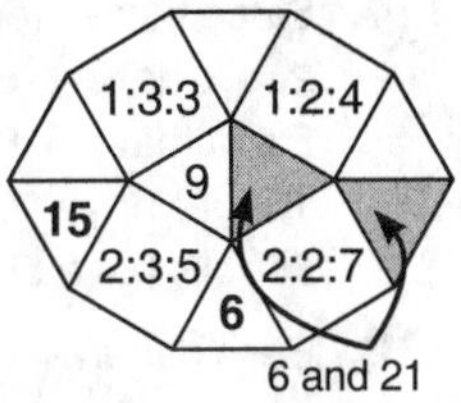
1:3:3
1:2:4
9
15
2:3:5
2:2:7
6
6 and 21

Suppose we place 21 in the middle-right triangle as shown. Then, the only ratio of three integers with 21 that simplifies to 1:2:4 is 21:[42]:[84] = 1:2:4.

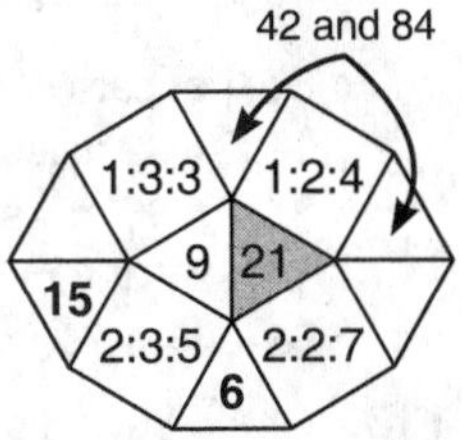
42 and 84
1:3:3
1:2:4
9
21
15
2:3:5
2:2:7
6

However, the top triangle also touches the top-left square. We cannot make a ratio of integers with 9 and either 42 or 84 that simplifies to 1:3:3. ✖

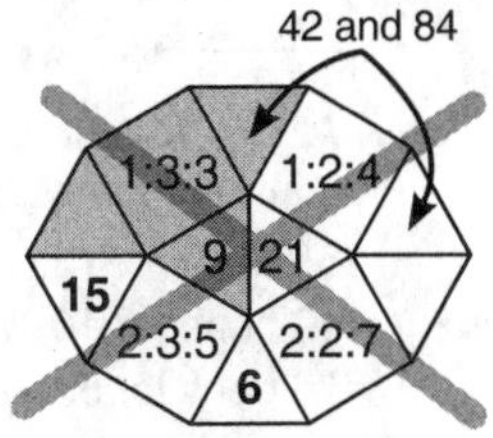
42 and 84
1:3:3
1:2:4
9
21
15
2:3:5
2:2:7
6

Therefore, we place the 6 and 21 in the triangles around the bottom-right square as shown.

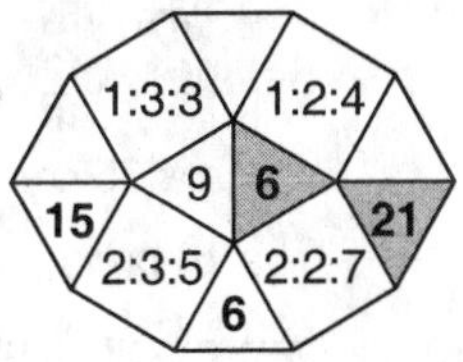
1:3:3
1:2:4
9
6
21
15
2:3:5
2:2:7
6

The ratio in the top-right square is 1:2:4. Since 6 is a multiple of both 1 and 2 (but not 4), we have

6:[12]:[24] = 1:2:4, *or*

[3]:6:[12] = 1:2:4.

So, the two other triangles that touch this square are (12 and 24) or (3 and 12).

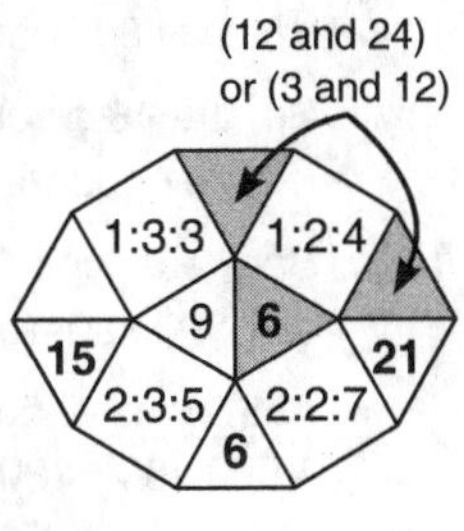
(12 and 24)
or (3 and 12)
1:3:3
1:2:4
9
6
21
15
2:3:5
2:2:7
6

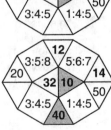

There is no ratio of three integers with 9 and either 12 or 24 that simplifies to 1:3:3.

So, only 3 can be in the triangle that touches both the top-left and top-right squares.

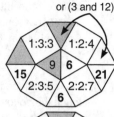

(12 and 24)
or (3 and 12)

Therefore, we place the 3 and 12 as shown.

Finally, 3:9:$\boxed{9}$ is the only ratio of three integers with 3 and 9 that simplifies to 1:3:3. So, the far-left triangle is 9.

111. Since there are 85 sheets per roll,
2 rolls have $2 \cdot 85 = $ **170** sheets,
5 rolls have $5 \cdot 85 = $ **425** sheets,
10 rolls have $10 \cdot 85 = $ **850** sheets, and
55 rolls have $55 \cdot 85 = $ **4,675** sheets.

112. Bess can travel 160 miles using 5 gallons. So, she travels $\frac{160 \text{ mi}}{5 \text{ gal}} = \frac{160}{5}$ miles per gallon = **32 miles per gallon**.
So, her car will go
$1 \cdot 32 = $ **32** miles with 1 gallon of gas,
$3 \cdot 32 = $ **96** miles with 3 gallons of gas,
$10 \cdot 32 = $ **320** miles with 10 gallons of gas, and
$55 \cdot 32 = $ **1,760** miles with 55 gallons of gas.

113. Grogg buys 6 pens for $2.10 = 210$ cents. So, he spends
$\frac{210 \text{ cents}}{6 \text{ pens}} = \frac{210}{6}$ cents per pen = **35 cents per pen**.
At this rate, he can buy
1 pen for $1 \cdot 35 = $ **35** cents (**$0.35**),
5 pens for $5 \cdot 35 = $ **175** cents (**$1.75**),
10 pens for $10 \cdot 35 = $ **350** cents (**$3.50**), and
55 pens for $55 \cdot 35 = $ **1,925** cents (**$19.25**).

114. The 16-pound bag from Arthur's Grocery costs
$\frac{12 \text{ dollars}}{16 \text{ pounds}} = \frac{12}{16}$ dollars per pound $= \frac{3}{4}$ dollars per pound,
which is 75 cents per pound.

The 10-pound bag from Betty's Grocery costs
$\frac{8 \text{ dollars}}{10 \text{ pounds}} = \frac{8}{10}$ dollars per pound $= \frac{4}{5}$ dollars per pound,
which is 80 cents per pound.

So, **the 16-pound bag from Arthur's Grocery** costs less per pound.

115. Benji can type 1,500 words in 30 minutes, which is a rate of $1,500 \div 30 = $ **50** words per minute.

At the rate of 50 words per minute, Benji needs $5,000 \div 50 = $ **100** minutes to type his 5,000-word essay.

116. Forty-five baby dragons were born in the last 9 days, which is a rate of $45 \div 9 = $ **5** baby dragons born per day.

At the rate of 5 baby dragons per day, $5 \cdot 12 = $ **60** baby dragons will be born in the next 12 days on Beast Island.

117. a. $\frac{48 \text{ apples}}{8 \text{ pies}} = \frac{48}{8}$ apples per pie = **6 apples per pie**.

b. $\frac{24 \text{ ounces of sugar}}{8 \text{ pies}} = \frac{24}{8}$ ounces of sugar per pie
= **3 ounces of sugar per pie**.

c. $\frac{18 \text{ dollars}}{15 \text{ pounds of apples}} = \frac{6}{5}$ dollars per pound of apples.
$\frac{6}{5} = 1\frac{1}{5}$ and $\frac{1}{5} = \frac{20}{100}$, so $1\frac{1}{5}$ dollars per pound of apples is **1.20** dollars per pound of apples, or **$1.20** per pound of apples.

d. $\frac{15 \text{ pounds of apples}}{8 \text{ pies}} = \frac{15}{8}$ pounds of apples per pie
$= 1\frac{7}{8}$ **pounds of apples per pie**.

e. $\frac{24 \text{ ounces of sugar}}{48 \text{ apples}} = \frac{24}{48}$ ounces of sugar per apple
$= \frac{1}{2}$ **ounces of sugar per apple**.

f. $\frac{15 \text{ pounds}}{48 \text{ apples}} = \frac{15}{48}$ pounds per apple
$= \frac{5}{16}$ **pounds per apple**.

118. He receives $\frac{300 \text{ Torts}}{75 \text{ Beastbucks}} = \frac{300}{75}$ Torts per Beastbuck
= **4 Torts per Beastbuck**.

119. Since he will receive 4 Torts per Beastbuck, he will receive $4 \cdot 100 = $ **400** Torts for 100 Beastbucks.

120. He receives $\frac{75 \text{ Beastbucks}}{300 \text{ Torts}} = \frac{75}{300}$ Beastbucks per Tort
$= \frac{1}{4}$ **Beastbucks per Tort**.

121. Since Mr. Jones receives $\frac{1}{4}$ Beastbucks per Tort, he will receive $\frac{1}{4} \cdot 480 = $ **120** Beastbucks for 480 Torts.

— *or* —

Since 4 Torts are worth 1 Beastbuck, he will receive $480 \div 4 = $ **120** Beastbucks for 480 Torts.

122. The blimp traveled 90 miles in 3 hours, so its average speed was

$$\frac{90 \text{ miles}}{3 \text{ hours}} = \frac{30 \text{ miles}}{1 \text{ hour}} = \textbf{30 miles per hour}.$$

123. One day is 24 hours. So, in one hour, kudzu grows

$$\frac{15 \text{ in}}{1 \text{ day}} = \frac{15 \text{ in}}{24 \text{ hr}} = \frac{5 \text{ in}}{8 \text{ hr}} = \frac{5}{8} \textbf{ inches per hour}.$$

124. The speed of the ball was $\frac{145 \text{ ft}}{2\frac{1}{2} \text{ sec}}$.

To write the ball's speed as a number of feet per second, we first write this speed as the relationship between a whole number of feet and a whole number of seconds.

$$\frac{145 \text{ ft}}{2\frac{1}{2} \text{ sec}} \overset{\cdot 2}{\underset{\cdot 2}{=}} \frac{290 \text{ ft}}{5 \text{ sec}}$$

So, the average speed of the ball was

$$\frac{290 \text{ ft}}{5 \text{ sec}} = \frac{58 \text{ ft}}{1 \text{ sec}} = \textbf{58 feet per second}.$$

— *or* —

Since the ball traveled 145 feet in $2\frac{1}{2}$ seconds, it can travel $145 \div 2\frac{1}{2}$ feet in 1 second.

$$145 \div 2\frac{1}{2} = 145 \div \frac{5}{2} = 145 \cdot \frac{2}{5} = 58.$$

So, the average speed of the ball was **58 feet per second**.

125. Robbie's average speed is $\frac{15 \text{ mi}}{1\frac{1}{3} \text{ hr}}$.

To write this speed as a number of miles per hour, we first write this speed as the relationship between a whole number of miles and a whole number of hours.

$$\frac{15 \text{ mi}}{1\frac{2}{3} \text{ hr}} \overset{\cdot 3}{\underset{\cdot 3}{=}} \frac{45 \text{ mi}}{5 \text{ hr}}$$

So, Robbie's average speed is

$$\frac{45 \text{ mi}}{5 \text{ hr}} = \frac{9 \text{ mi}}{1 \text{ hr}} = \textbf{9 miles per hour}.$$

— *or* —

Since Robbie ran 15 miles in $1\frac{2}{3}$ hours, he can run $15 \div 1\frac{2}{3}$ miles in 1 hour.

$$15 \div 1\frac{2}{3} = 15 \div \frac{5}{3} = 15 \cdot \frac{3}{5} = 9.$$

So, Robbie's average speed is **9 miles per hour**.

126. Erica can run $4\frac{1}{2}$ meters per second. So, in 45 seconds, she can run $45 \cdot 4\frac{1}{2} = 45 \cdot \frac{9}{2} = \frac{405}{2} = \textbf{202}\frac{1}{2}$ **meters**.

127. The boat could travel 25 miles in one hour at maximum speed. In $\frac{1}{3}$ of an hour, the boat can only go $\frac{1}{3}$ as far: $\frac{1}{3} \cdot 25 = \frac{25}{3} = \textbf{8}\frac{1}{3}$ **miles**.

128. Grogg hikes a total of $8+8 = 16$ kilometers in $3+2 = 5$ hours. So, Grogg's average speed for the entire trip is

$$\frac{16 \text{ km}}{5 \text{ hr}} = \frac{16}{5} \text{ km per hour} = \textbf{3}\frac{1}{5} \textbf{ km per hour}.$$

129. Allison's team rowed a total of $5+3 = 8$ miles in $\frac{3}{4} + \frac{7}{12} = \frac{16}{12} = \frac{4}{3} = 1\frac{1}{3}$ hours.

So, their average speed was

$$\frac{8 \text{ mi}}{1\frac{1}{3} \text{ hr}} = \frac{24 \text{ mi}}{4 \text{ hr}} = \frac{6 \text{ mi}}{1 \text{ hr}} = \textbf{6 miles per hour}.$$

130. Four hours is 240 minutes, so Annalise walks 10 miles in 240 minutes. Therefore, she walks each mile in $240 \div 10 = \textbf{24}$ minutes.

— *or* —

It takes Annalise 4 hours to walk 10 miles. She will travel one tenth of the distance in one tenth of the time. So, it will take her $\frac{4}{10} = \frac{2}{5}$ hours to walk $\frac{10}{10} = 1$ mile.

One hour is 60 minutes, so $\frac{2}{5}$ hours is $\frac{2}{5} \cdot 60 = \textbf{24}$ minutes.

131. One hour is 60 minutes, so Urlich can paddle 5 miles in 60 minutes. Therefore, he paddles each mile in $60 \div 5 = 12$ minutes.

So, his $4\frac{1}{2}$-mile trip will take $4\frac{1}{2} \cdot 12 = \frac{9}{2} \cdot 12 = \textbf{54}$ minutes.

132. a. 5 mowers can mow 4 fields in 60 minutes. It will take 1 mower five times as long to mow the same number of fields.

So, 1 mower can mow 4 fields in $5 \cdot 60 = \textbf{300}$ minutes.

b. 1 mower can mow 4 fields in 300 minutes. It will take $\frac{1}{4}$ as long to mow 1 field.

So, 1 mower can mow 1 field in $\frac{1}{4} \cdot 300 = \textbf{75}$ minutes.

c. 1 mower can mow 1 field in 75 minutes. It will take 3 mowers $\frac{1}{3}$ as long to mow 1 field together.

So, 3 mowers can mow 1 field in $\frac{1}{3} \cdot 75 = \textbf{25}$ minutes.

d. 3 mowers can mow 1 field in 25 minutes. It will take them 5 times as long to mow 5 fields.

So, 3 mowers can mow 5 fields in $5 \cdot 25 = \textbf{125}$ minutes.

133. a. 6 hoses can fill 5 buckets in 15 minutes. It will take them $\frac{1}{5}$ as long to fill 1 bucket.

So, 6 hoses can fill 1 bucket in $\frac{1}{5} \cdot 15 = \textbf{3}$ minutes.

b. 6 hoses can fill 1 bucket in 3 minutes. So, in $24 = 8 \cdot 3$ minutes, they can fill 8 times as many buckets.

So, 6 hoses can fill $8 \cdot 1 = \textbf{8}$ buckets in 24 minutes.

c. 6 hoses can fill 8 buckets in 24 minutes. Three hoses will fill half as many buckets in the same time.

So, 3 hoses can fill $\frac{1}{2} \cdot 8 = \textbf{4}$ buckets in 24 minutes.

134. a. 3 gazellephants can eat 3 buckets in 3 minutes.

3 gazellephants can eat 6 buckets in twice the time it takes them to eat 3 buckets. So, 3 gazellephants can eat 6 buckets in $2 \cdot 3 = 6$ minutes.

6 gazellephants can eat 6 buckets in one-half the time it takes 3 gazellephants. So, 6 gazellephants can eat 6 buckets of peanuts in $\frac{1}{2} \cdot 6 = \textbf{3}$ minutes.

b. In part (a), we determined that 6 gazellephants eat 6 buckets in 3 minutes.

6 gazellephants will eat twice as many buckets in twice the time. So, 6 gazellephants can eat $2 \cdot 6 = \textbf{12}$ buckets in 6 minutes.

135. a. 12 paint-bots can paint 20 houses in 9 hours.

Painting 5 houses is $\frac{5}{20} = \frac{1}{4}$ the work of painting 20 houses. So, $\frac{1}{4}$ as many bots are needed to paint 5 houses as are needed to paint 20 houses. Therefore, $\frac{1}{4} \cdot 12 = \textbf{3}$ paint-bots can paint 5 houses in 9 hours.

b. 3 paint-bots can paint 5 houses in 9 hours.

9 paint-bots can paint 5 houses in one-third the time it takes 3 paint-bots. Therefore, 9 paint-bots can paint 5 houses in $\frac{1}{3} \cdot 9 = 3$ hours.

9 paint-bots can paint twice as many houses in 6 hours as they can in 3 hours. So, 9 paint-bots can paint $5 \cdot 2 = $ **10** houses in 6 hours.

— *or* —

12 paint-bots can paint 20 houses in 9 hours.

9 paint-bots can paint $\frac{9}{12} = \frac{3}{4}$ as many houses as 12 paint-bots.

In 6 hours, only $\frac{6}{9} = \frac{2}{3}$ as many houses can be painted as can be painted in 9 hours.

So, 9 paint-bots can paint $\frac{3}{4} \cdot \frac{2}{3} = \frac{1}{2}$ as many houses in 6 hours as 12 paint bots in 9 hours.

Therefore, 9 paint-bots can paint $\frac{1}{2} \cdot 20 = $ **10** houses in 6 hours.

136. a. 10 carpenters take 18 days to build 8 sheds.

Building 12 sheds is $\frac{12}{8} = 1\frac{1}{2}$ times the work of building 8 sheds. So, it takes 10 carpenters $1\frac{1}{2}$ times as long to build 12 sheds as it takes them to build 8 sheds.

Therefore, 10 carpenters take $18 \cdot 1\frac{1}{2} = $ **27** days to build 12 sheds.

b. 10 carpenters take 18 days to build 8 sheds.

1 carpenter takes 10 times as long to build 8 sheds as it takes 10 carpenters. Therefore, 1 carpenter takes $18 \cdot 10 = 180$ days to build 8 sheds.

3 carpenters take one third as long to build 8 sheds as it takes 1 carpenter. Therefore, 3 carpenters take $\frac{1}{3} \cdot 180 = 60$ days to build 8 sheds.

15 days is $\frac{15}{60} = \frac{1}{4}$ as long as 60 days. So, 3 carpenters building for 15 days can only build $\frac{1}{4}$ as many sheds as they can in 60 days.

Therefore, 3 carpenters working for 15 days can build $8 \cdot \frac{1}{4} = $ **2** sheds.

— *or* —

10 carpenters take 18 days to build 8 sheds.

3 carpenters can build $\frac{3}{10}$ as many sheds as 10 carpenters can.

In 15 days, only $\frac{15}{18} = \frac{5}{6}$ as many sheds can be built as can be built in 18 days.

So, 3 carpenters can build $\frac{3}{10} \cdot \frac{5}{6} = \frac{1}{4}$ as many sheds in 15 days as 10 carpenters can build in 18 days.

Therefore, 3 carpenters working for 15 days can build $8 \cdot \frac{1}{4} = $ **2** sheds.

137. a. 6 lumberjacks can chop 5 logs in 40 minutes.

24 lumberjacks can chop 5 logs in one fourth the time it takes 6 lumberjacks to chop the logs. Therefore, 24 lumberjacks can chop 5 logs in $\frac{1}{4} \cdot 40 = 10$ minutes.

So, 24 lumberjacks will chop 3 times as much wood in 30 minutes as they can in 10 minutes. Therefore, 24 lumberjacks can chop $3 \cdot 5 = $ **15** logs in 30 minutes.

b. We begin by considering how long it will take for 1 lumberjack to chop 1 log.

6 lumberjacks can chop 5 logs in 40 minutes.

1 lumberjack takes six times as long as 6 lumberjacks to chop 5 logs. So, 1 lumberjack can chop 5 logs in $6 \cdot 40 = 240$ minutes.

1 lumberjack can chop 1 log in one fifth the time it takes him or her to chop 5 logs. So, 1 lumberjack can chop 1 log in $\frac{1}{5} \cdot 240 = 48$ minutes.

Now, we use our 1-lumberjack-1-log time to find how many lumberjacks can chop 6 logs in 72 minutes.

1 lumberjack takes six times as long to chop 6 logs as it takes him or her to chop 1 log. So, 1 lumberjack can chop 6 logs in $6 \cdot 48 = 288$ minutes.

72 minutes is one fourth as long as 288 minutes. So, it will take 4 times as many lumberjacks to chop 6 logs in 72 minutes as it takes to chop the logs in 288 minutes. Therefore, it takes $4 \cdot 1 = $ **4** lumberjacks to chop 6 logs in 72 minutes.

— *or* —

6 lumberjacks can chop 5 logs in 40 minutes.

72 minutes is $\frac{72}{40} = \frac{9}{5}$ of 40 minutes. So, 6 lumberjacks can chop $\frac{9}{5}$ as many logs in 72 minutes as they can chop in 40 minutes. Therefore, 6 lumberjacks can chop $\frac{9}{5} \cdot 5 = 9$ logs in 72 minutes.

6 logs is $\frac{6}{9} = \frac{2}{3}$ of 9 logs. So, it takes $\frac{2}{3}$ as many lumberjacks to chop 6 logs as it takes to chop 9 logs. Therefore, $\frac{2}{3} \cdot 6 = $ **4** lumberjacks can chop 6 logs in 72 minutes.

RATIOS & RATES
Conversions   64-70

138. There are $3 \cdot 5{,}280 = $ **15,840** feet in 3 miles.

139. Twenty yards is $20 \cdot 3 = 60$ feet, and 60 feet is $60 \cdot 12 = $ **720** inches.

140. Thirty kilometers is $30 \cdot 1{,}000 = 30{,}000$ meters, and 30,000 meters is $30{,}000 \cdot 100 = $ **3,000,000** centimeters.

141. There are 7 days in 1 week, 24 hours in 1 day, and 60 minutes in 1 hour.

One week is 7 days, and 7 days is $7 \cdot 24 = 168$ hours, and 168 hours is $168 \cdot 60 = $ **10,080** minutes.

142. There are 60 minutes in an hour and 60 seconds in a minute, so one hour is $60 \cdot 60 = 3{,}600$ seconds.

So, a person jogging 2 meters per second will jog $2 \cdot 3{,}600 = 7{,}200$ meters in an hour.

Since there are 1,000 meters in a kilometer, jogging 7,200 meters in an hour is equal to jogging

$$\frac{7,200}{1,000} = \frac{72}{10} = \textbf{7.2} \text{ or } \textbf{7}\frac{1}{5} \text{ kilometers in an hour.}$$

143. The cups units cancel, and $15 \cdot 8 = 120$, so we have

$$15 \text{ cups} = 15 \text{ cups} \cdot \frac{8 \text{ fluid ounces}}{1 \text{ cup}} = \textbf{120} \text{ fluid ounces.}$$

144. The ounces units cancel, and $\frac{60}{16} = \frac{15}{4} = 3\frac{3}{4}$, so we have

$$60 \text{ ounces} = 60 \text{ ounces} \cdot \frac{1 \text{ pound}}{16 \text{ ounces}} = \textbf{3}\frac{3}{4} \text{ pounds.}$$

145. The minutes units cancel, and $\frac{150}{60} = \frac{5}{2} = 2\frac{1}{2}$, so we have

$$150 \text{ ft per min} = \frac{150 \text{ feet}}{1 \text{ minute}} \cdot \frac{1 \text{ minute}}{60 \text{ seconds}} = \textbf{2}\frac{1}{2} \text{ ft per sec.}$$

146. To convert from nanoseconds to shakes, we use the conversion factor $\frac{1 \text{ shake}}{10 \text{ nanoseconds}}$ so that the nanoseconds units cancel.

$$6 \text{ nanoseconds} = 6 \text{ nanoseconds} \cdot \frac{1 \text{ shake}}{10 \text{ nanoseconds}} = \textbf{}\frac{3}{5} \text{ shakes}$$

147. To convert from fathoms to feet, we use the conversion factor $\frac{6 \text{ feet}}{1 \text{ fathom}}$ so that the fathoms units cancel.

$$30 \text{ fathoms} = 30 \text{ fathoms} \cdot \frac{6 \text{ feet}}{1 \text{ fathom}} = \textbf{180} \text{ feet}$$

148. To convert from nautical miles to meters, we use the conversion factor $\frac{1,852 \text{ meters}}{1 \text{ nautical mile}}$ so that the nautical miles units cancel.

$$2 \text{ nautical miles} = 2 \text{ nautical miles} \cdot \frac{1,852 \text{ meters}}{1 \text{ nautical mile}} = \textbf{3,704} \text{ m}$$

149. To convert from inches to cubits, we use the conversion factor $\frac{1 \text{ cubit}}{18 \text{ inches}}$ so that the inches units cancel.

$$90 \text{ inches} = 90 \text{ inches} \cdot \frac{1 \text{ cubit}}{18 \text{ inches}} = \textbf{5} \text{ cubits}$$

150. To convert from joules to kilocalories, we use the conversion factor $\frac{1 \text{ kilocalorie}}{4,184 \text{ joules}}$ so that the joules units cancel.

$$10 \text{ joules} = 10 \text{ joules} \cdot \frac{1 \text{ kilocalorie}}{4,184 \text{ joules}} = \textbf{}\frac{5}{2,092} \text{ kilocalories}$$

151. To convert from drams per fortnight to ounces per fortnight, we use the conversion factor $\frac{1 \text{ ounce}}{16 \text{ drams}}$ so that the drams units cancel.

$$6 \text{ drams per fortnight} = \frac{6 \text{ drams}}{1 \text{ fortnight}} \cdot \frac{1 \text{ ounce}}{16 \text{ drams}} = \textbf{}\frac{3}{8} \text{ oz per fortnight}$$

152. To convert from bytes per jiffy to bytes per second, we use the conversion factor $\frac{100 \text{ jiffies}}{1 \text{ second}}$ so that the jiffies units cancel.

$$200 \text{ bytes per jiffy} = \frac{200 \text{ bytes}}{1 \text{ jiffy}} \cdot \frac{100 \text{ jiffies}}{1 \text{ second}} = \textbf{20,000} \text{ bytes per sec}$$

153. To convert from dollars per meter to cents per meter, we use the conversion factor $\frac{100 \text{ cents}}{1 \text{ dollar}}$ so that the dollars units cancel.

We show how these units cancel when we use this conversion factor in Problem 161.

154. To convert from calories per gram to calories per kilogram, we use the conversion factor $\frac{1,000 \text{ g}}{1 \text{ kg}}$ so that the grams units cancel.

$$5 \text{ calories per g} = \frac{5 \text{ calories}}{1 \text{ g}} \cdot \frac{1,000 \text{ g}}{1 \text{ kg}} = \textbf{5,000} \text{ calories per kg}$$

155. To convert from feet per minute to miles per minute, we use the conversion factor $\frac{1 \text{ mi}}{5,280 \text{ ft}}$ so that the feet units cancel.

We show how these units cancel when we use this conversion factor in Problem 162.

156. To convert from cents per meter to dollars per meter, we use the conversion factor $\frac{1 \text{ dollar}}{100 \text{ cents}}$ so that the cents units cancel.

$$10 \text{ cents per meter} = \frac{10 \text{ cents}}{1 \text{ meter}} \cdot \frac{1 \text{ dollar}}{100 \text{ cents}} = \textbf{}\frac{1}{10} \text{ dollars per meter}$$

157. To convert from grains per kilogram to grains per gram, we use the conversion factor $\frac{1 \text{ kg}}{1,000 \text{ g}}$ so that the kilograms units cancel.

We show how these units cancel when we use this conversion factor in Problem 159.

158. To convert from gallons per foot to gallons per mile, we use the conversion factor $\frac{5,280 \text{ ft}}{1 \text{ mi}}$ so that the feet units cancel.

We show how these units cancel when we use this conversion factor in problem 160.

Below is the correct matching for #153-158.

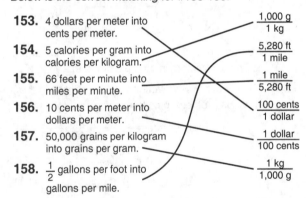

153. 4 dollars per meter into cents per meter.

154. 5 calories per gram into calories per kilogram.

155. 66 feet per minute into miles per minute.

156. 10 cents per meter into dollars per meter.

157. 50,000 grains per kilogram into grains per gram.

158. $\frac{1}{2}$ gallons per foot into gallons per mile.

$\frac{1,000 \text{ g}}{1 \text{ kg}}$

$\frac{5,280 \text{ ft}}{1 \text{ mile}}$

$\frac{1 \text{ mile}}{5,280 \text{ ft}}$

$\frac{100 \text{ cents}}{1 \text{ dollar}}$

$\frac{1 \text{ dollar}}{100 \text{ cents}}$

$\frac{1 \text{ kg}}{1,000 \text{ g}}$

159. We use the conversion factor $\frac{1 \text{ kg}}{1,000 \text{ g}}$ so that the kilograms units cancel.

$$50,000 \text{ grains per kg} = \frac{50,000 \text{ grains}}{1 \text{ kg}} \cdot \frac{1 \text{ kg}}{1,000 \text{ g}} = \textbf{50} \text{ grains per g}$$

160. We use the conversion factor $\frac{5,280 \text{ ft}}{1 \text{ mi}}$ so that the feet units cancel.

$$\frac{1}{2} \text{ gallons per foot} = \frac{\frac{1}{2} \text{ gal}}{1 \text{ ft}} \cdot \frac{5,280 \text{ ft}}{1 \text{ mi}} = \textbf{2,640} \text{ gallons per mile}$$

161. We use the conversion factor $\frac{100 \text{ cents}}{1 \text{ dollar}}$ so that the dollars units cancel.

$$4 \text{ dollars per meter} = \frac{4 \text{ dollars}}{1 \text{ meter}} \cdot \frac{100 \text{ cents}}{1 \text{ dollar}} = \textbf{400} \text{ cents per meter}$$

162. We use the conversion factor $\frac{1 \text{ mile}}{5{,}280 \text{ ft}}$ so that the feet units cancel.

$$66 \text{ feet per minute} = \frac{66 \text{ ft}}{1 \text{ min}} \cdot \frac{1 \text{ mi}}{5{,}280 \text{ ft}} = \frac{1}{80} \text{ miles per minute}$$

163. The gallons and minutes units cancel. We can also cancel some factors in the numerator and denominator.

$$30 \text{ gallons in 8 minutes} = \frac{30 \text{ gal}}{8 \text{ min}} \cdot \frac{128 \text{ fl oz}}{1 \text{ gal}} \cdot \frac{1 \text{ min}}{60 \text{ sec}}$$

$$= \frac{30 \cdot \overset{16}{128} \text{ fl oz}}{8 \cdot 60 \text{ sec}}$$

$$= \frac{30 \cdot 16 \text{ fl oz}}{\underset{2}{60} \text{ sec}}$$

$$= \mathbf{8} \text{ fluid ounces per second}$$

164. The cups and batches units cancel. We can also cancel some factors in the numerator and denominator.

$$3 \text{ cups for 2 batches} = \frac{3 \text{ cups}}{2 \text{ batches}} \cdot \frac{120 \text{ grams}}{1 \text{ cup}} \cdot \frac{1 \text{ batch}}{12 \text{ cookies}}$$

$$= \frac{3 \cdot \overset{10}{120} \text{ grams}}{2 \cdot 12 \text{ cookies}}$$

$$= \mathbf{15} \text{ grams per cookie}$$

165. To convert from feet per hour to inches per minute, we must cancel the feet and hour units.

$$200 \text{ feet per hour} = \frac{200 \text{ ft}}{1 \text{ hr}}.$$

So, we choose the conversion factors with "feet" in the denominator and "hours" in the numerator:

$$\frac{1 \text{ ft}}{12 \text{ in}} \qquad \boxed{\frac{12 \text{ in}}{1 \text{ ft}}} \qquad \boxed{\frac{1 \text{ hr}}{60 \text{ min}}} \qquad \frac{60 \text{ min}}{1 \text{ hr}}$$

We show how the units cancel when we use these two conversion factors in the next problem.

166. We use the two conversion factors chosen in the previous problem to compute

$$200 \text{ feet per hour} = \frac{200 \text{ ft}}{1 \text{ hr}} \cdot \frac{12 \text{ in}}{1 \text{ ft}} \cdot \frac{1 \text{ hr}}{60 \text{ min}}$$

$$= \frac{200 \cdot 12 \text{ in}}{\underset{5}{60} \text{ min}}$$

$$= \mathbf{40} \text{ inches per minute.}$$

167. To convert from gallons per day into fluid ounces per hour, we must cancel the gallons and day units.

$$450 \text{ gallons per day} = \frac{450 \text{ gal}}{1 \text{ day}}.$$

So, we choose the conversion factors with "gallons" in the denominator and "days" in the numerator:

$$\frac{1 \text{ gal}}{128 \text{ fl oz}} \qquad \boxed{\frac{128 \text{ fl oz}}{1 \text{ gal}}} \qquad \boxed{\frac{1 \text{ day}}{24 \text{ hr}}} \qquad \frac{24 \text{ hr}}{1 \text{ day}}$$

We show how the units cancel when we use these two conversion factors in the next problem.

168. We use the two conversion factors chosen in the previous problem to compute

$$450 \text{ gallons per day} = \frac{450 \text{ gal}}{1 \text{ day}} \cdot \frac{128 \text{ fl oz}}{1 \text{ gal}} \cdot \frac{1 \text{ day}}{24 \text{ hr}}$$

$$= \frac{450 \cdot \overset{16}{128} \text{ fl oz}}{\underset{3}{24} \text{ hr}}$$

$$= \mathbf{2{,}400} \text{ fluid ounces per hour.}$$

169. To convert from pounds per cup to ounces per teaspoon, we must cancel the pounds and cups units.

$$7 \text{ pounds per cup} = \frac{7 \text{ lbs}}{1 \text{ c}}.$$

So, we choose the conversion factors with "pounds" in the denominator and "cups" in the numerator:

$$\boxed{\frac{1 \text{ c}}{48 \text{ tsp}}} \qquad \frac{48 \text{ tsp}}{1 \text{ c}} \qquad \frac{1 \text{ lb}}{16 \text{ oz}} \qquad \boxed{\frac{16 \text{ oz}}{1 \text{ lb}}}$$

We show how the units cancel when we use these two conversion factors in the next problem.

170. We use the two conversion factors chosen in the previous problem to compute

$$7 \text{ pounds per cup} = \frac{7 \text{ lbs}}{1 \text{ c}} \cdot \frac{1 \text{ c}}{48 \text{ tsp}} \cdot \frac{16 \text{ oz}}{1 \text{ lb}}$$

$$= \frac{7 \cdot 16 \text{ oz}}{\underset{3}{48} \text{ tsp}}$$

$$= \frac{7}{3} \text{ ounces per teaspoon}$$

$$= 2\frac{1}{3} \text{ ounces per teaspoon.}$$

171. To convert from miles per hour to feet per minute, we must cancel the mile and hour units.

$$3 \text{ miles per hour} = \frac{3 \text{ mi}}{1 \text{ hr}}.$$

5,280 feet equals 1 mile. We use the conversion factor with "miles" in the denominator: $\frac{5{,}280 \text{ ft}}{1 \text{ mi}}$.

60 minutes equals 1 hour. We use the conversion factor with "hours" in the numerator: $\frac{1 \text{ hr}}{60 \text{ min}}$.

Therefore, we have

$$3 \text{ miles per hour} = \frac{3 \text{ mi}}{1 \text{ hr}} \cdot \frac{5{,}280 \text{ ft}}{1 \text{ mi}} \cdot \frac{1 \text{ hr}}{60 \text{ min}}$$

$$= \frac{3 \cdot \overset{88}{5{,}280} \text{ ft}}{60 \text{ min}}$$

$$= \mathbf{264} \text{ feet per minute.}$$

172. To convert from cents per inch to dollars per yard, we must cancel the cents and inch units.

$$15 \text{ cents per inch} = \frac{15 \text{ cents}}{1 \text{ in}}.$$

100 cents equals 1 dollar. We use the conversion factor with "cents" in the denominator: $\frac{1 \text{ dollar}}{100 \text{ cents}}$.

36 inches equals 1 yard. We use the conversion factor with "inches" in the numerator: $\frac{36 \text{ in}}{1 \text{ yd}}$.

Therefore, we have

$$15 \text{ cents per inch} = \frac{15 \text{ cents}}{1 \text{ in}} \cdot \frac{1 \text{ dollar}}{100 \text{ cents}} \cdot \frac{36 \text{ in}}{1 \text{ yd}}$$

$$= \frac{\overset{3}{15} \cdot 36 \text{ dollars}}{\underset{20}{100} \text{ yd}}$$

$$= \frac{3 \cdot \overset{9}{36} \text{ dollars}}{\underset{5}{20} \text{ yd}}$$

$$= \frac{27}{5} \text{ dollars per yard}$$

$$= \mathbf{5\frac{2}{5}} \text{ dollars per yard.}$$

$\frac{2}{5} = \frac{40}{100}$, so $5\frac{2}{5}$ dollars per yard is more commonly written as **$5.40** per yard. .

173. To convert from meters per second to kilometers per hour, we must cancel the meters and seconds units.

$$3 \text{ meters per second} = \frac{3 \text{ m}}{1 \text{ sec}}.$$

1,000 meters equals 1 kilometer. We use the conversion factor with "meters" in the denominator: $\frac{1 \text{ km}}{1{,}000 \text{ m}}$.

One minute is 60 seconds, and one hour is 60 minutes, so $60 \cdot 60 = 3{,}600$ seconds equals 1 hour. We use the conversion factor with "seconds" in the numerator: $\frac{3{,}600 \text{ sec}}{1 \text{ hr}}$.

Therefore, we have

$$3 \text{ meters per second} = \frac{3 \text{ m}}{1 \text{ sec}} \cdot \frac{1 \text{ km}}{1{,}000 \text{ m}} \cdot \frac{3{,}600 \text{ sec}}{1 \text{ hr}}$$

$$= \frac{3 \cdot \overset{18}{3{,}600} \text{ km}}{\underset{5}{1{,}000} \text{ hr}}$$

$$= \frac{54}{5} \text{ km per hour}$$

$$= 10\frac{4}{5} \text{ km per hour}.$$

— *or* —

To convert from seconds to hours, we could first convert from seconds to minutes, then from minutes to hours. To do this, we must choose two conversion factors so that all necessary units cancel: $\frac{60 \text{ sec}}{1 \text{ min}}$ and $\frac{60 \text{ min}}{1 \text{ hr}}$.

Therefore, we have

$$3 \text{ meters per second} = \frac{3 \text{ m}}{1 \text{ sec}} \cdot \frac{1 \text{ km}}{1{,}000 \text{ m}} \cdot \frac{60 \text{ sec}}{1 \text{ min}} \cdot \frac{60 \text{ min}}{1 \text{ hr}}$$

$$= \frac{3 \cdot 60 \cdot \overset{3}{60} \text{ km}}{\underset{50}{1{,}000} \text{ hr}}$$

$$= \frac{3 \cdot 60 \cdot \overset{6}{3} \text{ km}}{\underset{5}{50} \text{ hr}}$$

$$= \frac{54}{5} \text{ km per hour}$$

$$= 10\frac{4}{5} \text{ km per hour}$$

174. To convert from gallons per day to fluid ounces per minute, we must cancel the gallons and day units.

$$45 \text{ gallons per day} = \frac{45 \text{ gal}}{1 \text{ day}}.$$

128 fluid ounces equals 1 gallon. We use the conversion factor with "gallons" in the denominator: $\frac{128 \text{ fl oz}}{1 \text{ gal}}$.

One day is 24 hours, and one hour is 60 minutes, so $24 \cdot 60 = 1{,}440$ minutes equals 1 day. We use the conversion factor with "day" in the numerator: $\frac{1 \text{ day}}{1{,}440 \text{ min}}$.

Therefore, we have

$$45 \text{ gallons per day} = \frac{45 \text{ gal}}{1 \text{ day}} \cdot \frac{128 \text{ fl oz}}{1 \text{ gal}} \cdot \frac{1 \text{ day}}{1{,}440 \text{ min}}$$

$$= \frac{\overset{}{45} \cdot 128 \text{ fl oz}}{\underset{32}{1{,}440} \text{ min}}$$

$$= \textbf{4 fluid ounces per minute.}$$

— *or* —

To convert from days to minutes, we could first convert from days to hours, then from hours to minutes. To do this, we must choose two conversion factors so that all necessary units cancel: $\frac{1 \text{ day}}{24 \text{ hr}}$ and $\frac{1 \text{ hr}}{60 \text{ min}}$.

Therefore, we have

$$45 \text{ gallons per day} = \frac{45 \text{ gal}}{1 \text{ day}} \cdot \frac{128 \text{ fl oz}}{1 \text{ gal}} \cdot \frac{1 \text{ day}}{24 \text{ hr}} \cdot \frac{1 \text{ hr}}{60 \text{ min}}$$

$$= \frac{45 \cdot \overset{16}{128} \text{ fl oz}}{\underset{3}{24} \cdot 60 \text{ min}}$$

$$= \textbf{4 fluid ounces per minute.}$$

175. To convert from grams per serving to ounces per box, we must cancel the grams and serving units.

$$14 \text{ grams per serving} = \frac{14 \text{ g}}{1 \text{ serving}}.$$

One box is 8 servings, and we use the conversion factor with "servings" in the numerator: $\frac{8 \text{ servings}}{1 \text{ box}}$.

We estimate that 28 grams is 1 ounce. We use the conversion factor with "grams" in the denominator: $\frac{1 \text{ oz}}{28 \text{ g}}$.

Therefore, we have

$$14 \text{ grams per serving} = \frac{14 \text{ g}}{1 \text{ serving}} \cdot \frac{8 \text{ servings}}{1 \text{ box}} \cdot \frac{1 \text{ oz}}{28 \text{ g}}$$

$$= \frac{14 \cdot 8 \text{ oz}}{\underset{2}{28} \text{ box}}$$

$$= \textbf{4 ounces per box.}$$

176. To convert from centimeters per day to feet per week, we must cancel the centimeters and days units.

$$762 \text{ centimeters per day} = \frac{762 \text{ cm}}{1 \text{ day}}.$$

7 days equals 1 week. We use the conversion factor with "days" in the numerator: $\frac{7 \text{ days}}{1 \text{ week}}$.

To convert centimeters to feet, we first convert from centimeters to inches, then from inches to feet. 254 centimeters equals 100 inches. We use the conversion factor with "centimeters" in the denominator: $\frac{100 \text{ in}}{254 \text{ cm}}$.

12 inches equals 1 foot, and we need to cancel the "inches" unit of the conversion factor chosen above. So, we use the conversion factor with "inches" in the denominator: $\frac{1 \text{ ft}}{12 \text{ in}}$.

Therefore, we have

$$762 \text{ cm per day} = \frac{762 \text{ cm}}{1 \text{ day}} \cdot \frac{7 \text{ days}}{1 \text{ week}} \cdot \frac{100 \text{ in}}{254 \text{ cm}} \cdot \frac{1 \text{ ft}}{12 \text{ in}}$$

$$= \frac{\overset{3}{762} \cdot 7 \cdot \overset{25}{100} \text{ ft}}{\underset{}{254} \cdot \underset{3}{12} \text{ week}}$$

$$= \textbf{175 feet per week.}$$

177. 2 bears can catch 30 salmon in 54 minutes.

One bear can catch half as many salmon in 54 minutes as 2 bears can. So, 1 bear can catch $\frac{1}{2} \cdot 30 = 15$ salmon in 54 minutes.

One bear can catch one salmon in one fifteenth the amount of time it takes to catch 15 salmon. So, 1 bear can catch 1 salmon in $\frac{1}{15} \cdot 54 = \frac{18}{5} = 3\frac{3}{5}$ minutes.

Since 1 minute is 60 seconds, $3\frac{3}{5}$ minutes is $3\frac{3}{5} \cdot 60 = \textbf{216}$ seconds.

— *or* —

Since 1 minute is 60 seconds, 54 minutes is $54 \cdot 60 = 3{,}240$ seconds.

So, 2 bears can catch 30 salmon in 3,240 seconds.

One bear can catch half as many salmon in 3,240 seconds as 2 bears can. So, 1 bear can catch $\frac{1}{2} \cdot 30 = 15$ salmon in 3,240 seconds.

One bear can catch one salmon in one fifteenth the amount of time it takes to catch 15 salmon. So, 1 bear can catch 1 salmon in $\frac{1}{15} \cdot 3{,}240 = \textbf{216}$ seconds.

178. The ratio of Sarah's fruits to Jeremy's fruits is 4:5, and together they have 135 fruits. So, Sarah has $\frac{4}{9} \cdot 135 = 60$ fruits and Jeremy has $\frac{5}{9} \cdot 135 = 75$ fruits.

Sarah's ratio of lemons to limes is 2:3, and she has 60 fruits. So, Sarah has $\frac{2}{5} \cdot 60 = 24$ lemons and $\frac{3}{5} \cdot 60 = 36$ limes.

Jeremy's ratio of lemons to limes is 1:2, and he has 75 fruits. So, Jeremy has $\frac{1}{3} \cdot 75 = 25$ lemons and $\frac{2}{3} \cdot 75 = 50$ limes.

When they combine their fruits into the same basket, there are $24 + 25 = 49$ lemons and $36 + 50 = 86$ limes. Therefore, the ratio of lemons to limes is **49:86**.

179. The ratio tells us that $\frac{3}{12} = \frac{1}{4}$ of the perimeter comes from the shortest side, $\frac{4}{12} = \frac{1}{3}$ of the perimeter comes from the medium side, and $\frac{5}{12}$ of the perimeter comes from the longest side.

The perimeter is 60 inches, so the lengths of the three sides of the triangle are $\frac{1}{4} \cdot 60 = 15$ inches, $\frac{1}{3} \cdot 60 = 20$ inches, and $\frac{5}{12} \cdot 60 = 25$ inches.

The legs of a right triangle are its two shortest sides, and the area of a right triangle is equal to half the product of the lengths of its legs. Therefore, the area of the triangle is $\frac{15 \cdot 20}{2} = \textbf{150}$ sq in.

— *or* —

The ratio of the lengths of the three legs is 3:4:5. So, for some number of inches x, the shortest side is $3x$ inches long, the medium side is $4x$ inches long, and the longest side is $5x$ inches long.

Since the perimeter of the triangle is 60 inches, we have
$$3x + 4x + 5x = 60$$
$$12x = 60.$$

So, $x = \frac{60}{12} = 5$. The lengths of the triangle sides are $3x = 3 \cdot 5 = 15$ inches, $4x = 4 \cdot 5 = 20$ inches, and $5x = 5 \cdot 5 = 25$ inches.

The legs of a right triangle are its two shortest sides, and the area of a right triangle is equal to half the product of the lengths of its legs. Therefore, the area of the triangle is $\frac{15 \cdot 20}{2} = \textbf{150}$ sq in.

180. Since the tarantulemurs are crawling directly towards each other, the tarantulemurs move $2 + 3 = 5$ meters closer to each other each second. When they meet, they will have crawled a combined distance of 70 meters.

Crawling 70 meters at 5 meters per second takes $70 \div 5 = \textbf{14}$ seconds.

181. The ratio of Jorble's height to Yorble's height is 5:9.

So, for some value of x, Jorble is $5x$ inches tall and Yorble is $9x$ inches tall. The difference between their heights is $9x - 5x = 4x$ inches.

We are told that Jorble is 10 inches shorter than Yorble, so we have $4x = 10$. Solving for x, we get $x = \frac{10}{4} = \frac{5}{2}$. Therefore, Yorble is $9 \cdot \frac{5}{2} = \frac{45}{2} = \textbf{22}\frac{1}{2}$ inches tall, and Jorble is $5 \cdot \frac{5}{2} = \frac{25}{2} = 12\frac{1}{2}$ inches tall.

Check: The ratio of Jorble's height to Yorble's height is $12\frac{1}{2}:22\frac{1}{2} = 25:45 = 5:9$. ✓

182. The ratio of the height of the rectangle to its perimeter is 2:19. So, for some unit of length x, the rectangle's perimeter is $19x$ units, and its height is $2x$ units.

$2x$ ▭ $2x$
Perimeter = $19x$

Since the perimeter of a rectangle is the sum of its four side lengths, the sum of the two unknown sides is $19x - 2x - 2x = 15x$ units.

Therefore, the width of the rectangle is $\frac{15}{2}x$ units.

$\frac{15}{2}x$
$2x$ ▭ $2x$
$\frac{15}{2}x$

So, the ratio of the rectangle's height to its width is $2x:\frac{15}{2}x = 4x:15x$. We divide both quantities by x to simplify this to **4:15**.

183. We start by considering a simple example. Since the ratio of Alyssa's biking speed to her running speed is 5:2, we imagine that Alyssa bikes 5 mph and runs 2 mph.

If Alyssa's school is 20 miles away, it will take her $20 \div 5 = 4$ hours to bike to school, and $20 \div 2 = 10$ hours to run to school. So, the ratio of her biking time to running time is 4:10 = 2:5.

There is nothing special about the numbers or units that we used in our example! Alyssa can bike any distance 5 times in the same amount of time it takes her to run the same distance 2 times.

If x is the time it takes for Alyssa to bike to school 5 times or run to school 2 times, then the time it takes Alyssa to bike to school once is $\frac{1}{5}x$, and the time it takes Alyssa to run to school once is $\frac{1}{2}x$.

So, the ratio of the time it takes her to bike to school to the time it takes her to run to school is $\frac{1}{5}x$ to $\frac{1}{2}x$.

Dividing both quantities by x gives $\frac{1}{5}x:\frac{1}{2}x = \frac{1}{5}:\frac{1}{2}$. Multiplying both numbers by the least common multiple of their denominators, 10, gives $\frac{1}{5}:\frac{1}{2} = \textbf{2:5}$.

— *or* —

The faster Alyssa goes, the less time it will take her to get to school.

For example, if Alyssa travels twice as fast when biking than when running, it will take her half as long to get to school biking than it will running.

In other words, if the ratio of her biking speed to her running speed is 2:1, then the ratio of the time it takes to bike to school to the time it takes to run to school is 1:2.

Similarly, if Alyssa bikes $\frac{5}{2}$ as fast as she runs, it will take her $\frac{2}{5}$ as long to get to school.

In other words, if the ratio of her biking speed to her running speed is 5:2, then the ratio of the time it takes to bike to school to the time it takes to run to school is **2:5**.

184. When 15 blue cars leave the lot, the ratio of blue cars to red cars changes from 2:3 to 1:2.

Since the number of red cars in the parking lot does not change, we write our ratios using the same number of red cars. Since 6 is a multiple of 3 and of 2, we can write each ratio as a ratio of blue cars to 6 red cars.

<u>Before cars left:</u> 2:3 = 4:6

<u>After cars left:</u> 1:2 = 3:6

When 15 blue cars leave, the ratio of blue cars to red cars changes from 4:6 to 3:6. This tells us that, for each blue car that leaves the parking lot, there are 6 red cars in the lot. Fifteen blue cars left, so there are $15 \cdot 6 = $ **90** red cars.

— *or* —

The original ratio of the number of blue cars to red cars in the lot is 2:3. So, for some number c, there were $2c$ blue cars and $3c$ red cars in the lot.

After 15 blue cars exit the lot, $2c - 15$ blue cars and $3c$ red cars remain. Since the ratio of blue cars to red cars at that time is 1:2, we know there are twice as many red cars as blue cars remaining in the lot. So, $3c$ is twice as much as $2c - 15$. We write an equation to solve for c.

$$3c = 2 \cdot (2c - 15)$$
$$3c = 4c - 30$$
$$30 = 4c - 3c$$
$$30 = c$$

So, there are $3c = 3 \cdot 30 = $ **90** red cars in the lot.

Check: Since there are 90 red cars and 60:90 = 2:3, there were originally 60 blue cars in the lot. After 15 blue cars leave the lot, $60 - 15 = 45$ remain, and the ratio of blue cars to red cars is 45:90 = 1:2. ✓

185. The ratio of filled seats to empty seats is 4:1. So, $\frac{4}{5}$ of the seats in the auditorium are filled, and $\frac{1}{5}$ of the seats are empty.

The ratio of students to teachers is 5:1. So, $\frac{5}{6}$ of the filled seats hold students, and $\frac{1}{6}$ of the filled seats hold teachers.

$\frac{4}{5}$ of the seats in the auditorium are full, and teachers have $\frac{1}{6}$ of these filled seats. So, $\frac{4}{5} \cdot \frac{1}{6} = \frac{2}{15}$ of the auditorium seats are filled by teachers.

Since $\frac{1}{5}$ of all the seats are empty, the ratio of teachers to empty seats in the auditorium is $\frac{2}{15} : \frac{1}{5} = $ **2:3**.

— *or* —

We can create a diagram to represent the seats in the auditorium.

The ratio of students to teachers is 5:1. So, we can make rows of 6 filled seats, each with 5 students and 1 teacher. We use S for students, and T for teachers.

S S S S S T

The ratio of filled seats to empty seats is 4:1. So, we can make groups of 5 rows, each with 4 filled rows and 1 empty row. We use E in our diagram to represent an empty seat.

S S S S S T
S S S S S T
S S S S S T
S S S S S T
E E E E E

So, we can make groups of 30 auditorium seats, each with 20 filled by students, 4 filled by teachers, and 6 empty. Therefore, the ratio of teachers to empty seats is 4:6 = **2:3**.

186. It would be very complicated to think about how far Brody travels before each time he changes direction. However, we know his flying speed. If we know how long he was flying, then we can use his speed to find the total distance he flew.

Since each skateboard is traveling toward the other, the skateboards move $5 + 5 = 10$ feet closer to each other every second.

When they meet, they will have rolled a combined distance of 100 feet.

Rolling 100 feet at 10 feet per second takes $100 \div 10 = 10$ seconds.

Since Brody flies 7 feet per second for 10 seconds before the skateboards collide, he travels a total of $7 \cdot 10 = $ **70** feet.

Review
75-77

1. 0.9 has a 9 in the tenths place. So, $0.9 = \frac{9}{10}$.

2. 0.327 has a 3 in the tenths place, a 2 in the hundredths place, and a 7 in the thousandths place. So,

$$0.327 = \frac{3}{10} + \frac{2}{100} + \frac{7}{1,000}$$
$$= \frac{300}{1,000} + \frac{20}{1,000} + \frac{7}{1,000}$$
$$= \frac{327}{1,000}.$$

— *or* —

We can write any decimal with three digits to the right of the decimal point as a number of thousandths.

So, $0.327 = \frac{327}{1,000}$.

3. We can write any decimal with two digits to the right of the decimal point as a number of hundredths.

So, $0.46 = \frac{46}{100} = \frac{23}{50}$.

4. We can write any decimal with three digits to the right of the decimal point as a number of thousandths.

So, $6.128 = 6\frac{128}{1,000} = 6\frac{16}{125}$.

5. We have $27.014 = 27\frac{14}{1,000} = 27\frac{7}{500}$.

6. We have $505.0505 = 505\frac{505}{10,000} = 505\frac{101}{2,000}$.

7. $\frac{3}{10}$ is 3 tenths. So, to write $\frac{3}{10}$ as a decimal, we write a 3 in the tenths place: $\frac{3}{10} = \mathbf{0.3}$.

8. We have

$$\frac{53}{100} = \frac{50}{100} + \frac{3}{100}$$
$$= \frac{5}{10} + \frac{3}{100}.$$

So, to write $\frac{53}{100}$ as a decimal, we write 5 in the tenths place and 3 in the hundredths place: $\frac{53}{100} = \mathbf{0.53}$.

— *or* —

We can write any number of hundredths as a decimal with two digits to the right of the decimal point.

So, $\frac{53}{100} = \mathbf{0.53}$.

9. We have

$$\frac{809}{1,000} = \frac{800}{1,000} + \frac{0}{1,000} + \frac{9}{1,000}$$
$$= \frac{8}{10} + \frac{0}{100} + \frac{9}{1,000}.$$

So, to write $\frac{809}{1,000}$ as a decimal, we write 8 in the tenths place, 0 in the hundredths place, and 9 in the thousandths place: $\frac{809}{1,000} = \mathbf{0.809}$.

— *or* —

We can write any number of thousandths as a decimal with three digits to the right of the decimal point.

So, $\frac{809}{1,000} = \mathbf{0.809}$.

10. We begin by writing the fraction as a mixed number.

$$\frac{657}{100} = 6\frac{57}{100}.$$

Then, we can write any number of hundredths as a decimal with two digits to the right of the decimal point.

So, $\frac{57}{100} = 0.57$. Therefore, $\frac{657}{100} = 6\frac{57}{100} = \mathbf{6.57}$.

11. We have $\frac{2,021}{1,000} = 2\frac{21}{1,000} = \mathbf{2.021}$.

12. We have $23\frac{71}{10,000} = \mathbf{23.0071}$.

13. 1.43 is between 1.4 and 1.5. Since 1.43 is closer to 1.4 than to 1.5, we round 1.43 down to **1.4**.

14. 0.368 is between 0.3 and 0.4. Since 0.368 is closer to 0.4 than to 0.3, we round 0.368 up to **0.4**.

15. 77.7777 is between 77.77 and 77.78. Since 77.7777 is closer to 77.78 than to 77.77, we round 77.7777 up to **77.78**.

16. 1.245 is exactly halfway between 1.24 and 1.25. Numbers that are exactly in the middle are rounded up. So, we round 1.245 up to **1.25**.

17. 0.00317 is between 0.003 and 0.004. Since 0.00317 is closer to 0.003 than to 0.004, we round 0.00317 down to **0.003**.

18. 8.2497 is between 8.249 and 8.250. Since 8.2497 is closer to 8.250 than to 8.249, we round 8.2497 up to **8.250**.

19. For each number, we either place the decimal point

- two spaces to the left of a 6 that will remain a 6 when the number is rounded down, or

- two spaces to the left of a 5 that will become a 6 when the number is rounded up.

Below, we show the only way to place each decimal point *between two digits* so that the hundredths digit of each rounded number is 6.

19.659   5.6565   4565.464

Check: We round each answer to the nearest hundredth.
19.659 rounds to 19.6<u>6</u>. ✓
5.6565 rounds to 5.6<u>6</u>. ✓
4565.464 rounds to 4565.4<u>6</u>. ✓

20. Let x represent the result Grogg gets after rounding his decimal to the nearest hundredth.

When Grogg rounds x to the nearest tenth, he gets 0.4. The smallest number that rounds to 0.4 when rounded to the nearest tenth is 0.35.

So, the smallest result Grogg can get after rounding his decimal to the nearest hundredth is 0.35. The smallest number that rounds to 0.35 when rounded to the nearest hundredth is **0.345**.

Check: To the nearest hundredth, 0.345 rounds to 0.35. To the nearest tenth, 0.35 rounds to 0.4. Any number smaller than 0.345 rounds to 0.34 or less when rounded to the nearest hundredth, which cannot round to 0.4. ✓

21. We stack the numbers vertically so that matching place values are lined up. Then, we add the digits in each place value from right to left.

$$
\begin{array}{r} 2.35 \\ + 6.91 \\ \hline 6 \end{array}
\qquad
\begin{array}{r} {}^{1} \\ 2.35 \\ + 6.91 \\ \hline 26 \end{array}
\qquad
\begin{array}{r} {}^{1} \\ 2.36 \\ + 6.91 \\ \hline \mathbf{9.26} \end{array}
$$

22. We stack the numbers vertically so that matching place values are lined up, filling the empty hundredths place of 12.4 with 0. Then, we add the digits in each place value from right to left.

$$
\begin{array}{r} 12.40 \\ + 9.73 \\ \hline 3 \end{array}
\quad
\begin{array}{r} {}^{1} \\ 12.40 \\ + 9.73 \\ \hline 13 \end{array}
\quad
\begin{array}{r} {}^{11} \\ 12.40 \\ + 9.73 \\ \hline 2.13 \end{array}
\quad
\begin{array}{r} {}^{11} \\ 12.40 \\ + 9.73 \\ \hline \mathbf{22.13} \end{array}
$$

23. After stacking the numbers vertically with the place values aligned, we see that we cannot take 5 hundredths from 0 hundredths. So, we take 1 tenth from the tenths place of 5.90 and break it into 10 hundredths. This gives us 5 ones, 8 tenths, and 10 hundredths to subtract from.

$$
\begin{array}{r} {}^{8\ 10} \\ 5.9\!\!\!/0 \\ - 2.65 \\ \hline \end{array}
$$

Then, we subtract the digits in each place value from right to left.

$$
\begin{array}{r} {}^{8\ 10} \\ 5.9\!\!\!/0 \\ - 2.65 \\ \hline 5 \end{array}
\qquad
\begin{array}{r} {}^{8\ 10} \\ 5.9\!\!\!/0 \\ - 2.65 \\ \hline 25 \end{array}
\qquad
\begin{array}{r} {}^{8\ 10} \\ 5.9\!\!\!/0 \\ - 2.65 \\ \hline \mathbf{3.25} \end{array}
$$

24. We stack the numbers vertically and subtract as shown.

$$
\begin{array}{r} {}^{9} \\ {}^{0\ 18\ 10} \\ 10.1\!\!\!/0\!\!\!/0 \\ - 0.008 \\ \hline 2 \end{array}
\;
\begin{array}{r} {}^{9} \\ {}^{0\ 18\ 10} \\ 10.1\!\!\!/0\!\!\!/0 \\ - 0.008 \\ \hline 92 \end{array}
\;
\begin{array}{r} {}^{9} \\ {}^{0\ 18\ 10} \\ 10.1\!\!\!/0\!\!\!/0 \\ - 0.008 \\ \hline 092 \end{array}
\;
\begin{array}{r} {}^{9} \\ {}^{0\ 18\ 10} \\ 10.1\!\!\!/0\!\!\!/0 \\ - 0.008 \\ \hline 0.092 \end{array}
\;
\begin{array}{r} {}^{9} \\ {}^{0\ 18\ 10} \\ 10.1\!\!\!/0\!\!\!/0 \\ - 0.008 \\ \hline \mathbf{10.092} \end{array}
$$

25. To compute the perimeter, we add the three side lengths.

$$
\begin{array}{r} {}^{1} \\ 4.30 \\ 6.45 \\ + 9.90 \\ \hline 20.65 \end{array}
$$

So, the perimeter of the triangle is **20.65** cm.

26. Alice ran 6.35 miles out, then another 6.35 miles back. All together, she ran $6.35+6.35 = 12.7$ miles.

Ben ran 4.4 miles out, then another 4.4 miles back. All together, he ran $4.4+4.4 = 8.8$ miles.

So, Alice ran $12.7-8.8 = $ **3.9** more miles than Ben.

— *or* —

Alice ran $6.35-4.4 = 1.95$ miles farther than Ben did on the way out. Alice also ran 1.95 miles farther than Ben on the way back.

So, Alice ran $1.95+1.95 = $ **3.9** more miles than Ben.

27. We get the greatest possible difference by subtracting the smallest number from the largest number. It is easier to order these decimals if we write them vertically with their place values aligned.

The largest number in this list is 13.1. The smallest number in this list is 0.078.

So, the greatest possible difference that can be made by subtracting two of these numbers is $13.1-0.078 = $ **13.022**.

$$
\begin{array}{r}
13.1 \\
12.85 \\
3.54 \\
1.0065 \\
0.32 \\
0.23 \\
0.078
\end{array}
$$

Multiplying by 10   78-79

28. 0.8 is 8 tenths, or $\frac{8}{10}$. So, $0.8\times10 = \frac{8}{10}\times10 = \mathbf{8}$.

— *or* —

For any two adjacent place values, the place value on the left is ten times the place value to its right. So, multiplying a number by 10 moves each digit to the next-larger place value to its left.

$$
\begin{array}{cc} 1\text{'s} & \tfrac{1}{10}\text{'s} \\ 0 & .\ 8 \end{array} \times 10
$$
$$
= \ 8\ .
$$

This is the same as moving the *decimal point* one place to the *right*.

$$
0\ .\ 8 \times 10
$$
$$
= \ 0\ 8\ .
$$

So, $0.8\times10 = \mathbf{8}$.

29. 0.064 is 6 hundredths and 4 thousandths, or $\frac{6}{100}+\frac{4}{1,000}$. To multiply this sum by 10, we distribute the 10 as shown below.

$$
\begin{aligned}
0.064\times10 &= \left(\frac{6}{100}+\frac{4}{1,000}\right)\times10 \\
&= \left(\frac{6}{100}\times10\right)+\left(\frac{4}{1,000}\times10\right) \\
&= \frac{6}{10}+\frac{4}{100} \\
&= 0.64.
\end{aligned}
$$

So, $0.064\times10 = \mathbf{0.64}$.

— *or* —

Multiplying a number by 10 moves each digit to the next-larger place value to its left.

$$
\begin{array}{cccc} 1\text{'s} & \tfrac{1}{10}\text{'s} & \tfrac{1}{100}\text{'s} & \tfrac{1}{1,000}\text{'s} \\ 0 & .\ 0 & 6 & 4 \end{array} \times 10
$$
$$
= \ 0\ .\ 6\ 4
$$

This is the same as moving the *decimal point* one place to the *right*.

$$\underset{=\ \underline{0}\ \underline{0}\ .\ \underline{6}\ \underline{4}}{\underline{0}\ .\ \underline{0}\ \underline{6}\ \underline{4}\ \times 10}$$

So, $0.064 \times 10 =$ **0.64**.

30. Multiplying a number by 10 moves the number's decimal point one place to the right. So, $10 \times 3.7 =$ **37**.

31. Multiplying a number by 10 moves the number's decimal point one place to the right.

So, $0.901 \times 10 \times 10 = 9.01 \times 10 =$ **90.1**.

— *or* —

Multiplying a number by 10 once moves the number's decimal point one place to the right. So, multiplying a number by 10 *twice* moves the number's decimal point *two* places to the right.

So, $0.901 \times 10 \times 10 =$ **90.1**.

32. Since multiplication is associative and commutative, $10 \times 21.032 \times 10 = 21.032 \times 10 \times 10$. Multiplying a number by 10 twice moves the number's decimal point two places to the right.

So, $10 \times 21.032 \times 10 = 21.032 \times 10 \times 10 =$ **2,103.2**.

33. $10 \times 3.001 \times 10 = 3.001 \times 10 \times 10 =$ **300.1**.

34. Since $100 = 10 \times 10$, multiplying a number by 100 moves the number's decimal point two places to the right. Moving the decimal point in $12.3 = 12.30$ two places to the right gives 1,230.

So, $12.3 \times 100 =$ **1,230**.

35. Since $1,000 = 10 \times 10 \times 10$, multiplying a number by 1,000 moves the number's decimal point three places to the right. So, $1,000 \times 0.03405 =$ **34.05**.

36. Multiplying a number by 1,000 moves the number's decimal point three places to the right. So, $0.00002 \times 1,000 =$ **0.02**.

37. 10^5 is the product of five 10's. So, multiplying a number by 10^5 moves the number's decimal point five places to the right. Therefore, $0.040608 \times 10^5 =$ **4,060.8**.

38. To go from 1.821 to 18.21, we move the decimal point one place to the right. Multiplying a number by 10 moves the number's decimal point one place to the right.

So, $1.821 \times$ **10** $= 18.21$.

39. To go from 0.007 to 7, we move the decimal point three places to the right. Multiplying a number by $10 \times 10 \times 10 = 1,000$ moves the number's decimal point three places to the right.

So, $0.007 \times$ **1,000** $= 7$.

40. To go from 0.065 to 6.5, we move the decimal point two places to the right. Multiplying a number by $10 \times 10 = 100$ moves the number's decimal point two places to the right.

So, **100** $\times 0.065 = 6.5$.

41. When we multiply the missing number by 10, the decimal point moves one place to the *right*, giving 0.3. So, to find the missing number, we move the decimal point in 0.3 one place to the *left*. This gives 0.03.

So, **0.03** $\times 10 = 0.3$.

42. When we multiply the missing number by 100, the decimal point moves two places to the *right*, giving 71.3. So, to find the missing number, we move the decimal point in 71.3 two places to the *left*. This gives 0.713.

So, $100 \times$ **0.713** $= 71.3$.

43. When we multiply the missing number by 10,000, the decimal point moves four places to the *right*, giving 345. So, to find the missing number, we move the decimal point in 345 four places to the *left*. This gives 0.0345.

So, $10,000 \times$ **0.0345** $= 345$.

44. To go from 0.02689 to 268.9, we move the decimal point four places to the right. Multiplying a number by four 10's moves the number's decimal point four places to the right. Multiplying by four 10's is the same as multiplying by $10 \times 10 \times 10 \times 10 = 10^4$.

So, $0.02689 \times 10^4 = 268.9$. Therefore, $n =$ **4**.

45. To compute this quotient, we consider the relationship between multiplication and division.

If $0.043 \div 10 = \boxed{}$, then $\boxed{} \times 10 = 0.043$.

Since $0.0043 \times 10 = 0.043$, we have $0.043 \div 10 =$ **0.0043**.

Notice that dividing by 10 moves the decimal point one place to the left!

46. Multiplying 0.002 by 1,000 moves the decimal point in 0.002 three places to the right, giving the integer 2:

$$0.002 \times 1,000 = 2.$$

The only positive integer smaller than 2 is 1. Since 1 is half of 2, we can multiply 0.002 by half of 1,000 to get 1. Half of 1,000 is 500, so

$$0.002 \times 500 = 1.$$

Since 1 is the smallest positive integer, **500** is the smallest positive integer we can multiply by 0.002 to get an integer result.

47. Multiplying 0.0025 by 10,000 moves the decimal point in 0.0025 four places to the right, giving the integer 25:

$$0.0025 \times 10,000 = 25.$$

Both 10,000 and 25 are divisible by 25. So, multiplying 0.0025 by $\frac{1}{25}$ of 10,000 will give a result that is $\frac{1}{25}$ of 25. $\frac{1}{25}$ of 10,000 is $\frac{10,000}{25} = 400$, and $\frac{1}{25}$ of 25 is $\frac{25}{25} = 1$. So,

$$0.0025 \times 400 = 1.$$

Since 1 is the smallest positive integer, **400** is the smallest positive integer we can multiply by 0.0025 to get an integer result.

48. 1.5 is $1 + \frac{5}{10}$, and 0.1 is $\frac{1}{10}$. We use these fractions to compute 1.5×0.1 as shown below:

$$1.5 \times 0.1 = \left(1 + \frac{5}{10}\right) \times \frac{1}{10}$$
$$= \left(1 \times \frac{1}{10}\right) + \left(\frac{5}{10} \times \frac{1}{10}\right)$$
$$= \frac{1}{10} + \frac{5}{100}$$
$$= 0.15.$$

So, $1.5 \times 0.1 = \mathbf{0.15}$.

— or —

For any two adjacent place values, the place value on the right is one tenth the place value to its left. So, multiplying a number by 0.1 moves each digit to the next-smaller place value to its right.

This is the same as moving the *decimal point* one place to the *left*.

So, $1.5 \times 0.1 = \mathbf{0.15}$.

49. 0.09 is $\frac{9}{100}$, and 0.1 is $\frac{1}{10}$. We use these fractions to compute 0.1×0.09 as shown below:

$$0.1 \times 0.09 = \frac{1}{10} \times \frac{9}{100}$$
$$= \frac{9}{1,000}$$
$$= 0.009.$$

So, $0.1 \times 0.09 = \mathbf{0.009}$.

— or —

Multiplying a number by 0.1 moves each digit to the next-smaller place value to its right.

This is the same as moving the *decimal point* one place to the *left*.

So, $0.1 \times 0.09 = \mathbf{0.009}$.

50. Multiplying a number by 0.1 moves the number's decimal point one place to the left. So, $0.1 \times 8 = \mathbf{0.8}$.

51. Multiplying a number by 0.1 moves the number's decimal point one place to the left. So, $340 \times 0.1 = 34.0 = \mathbf{34}$.

52. Multiplying a number by 0.1 moves the number's decimal point one place to the left. So, $12.34 \times 0.1 = \mathbf{1.234}$.

53. Multiplying a number by 0.1 moves the number's decimal point one place to the left.

So, $6.5 \times 0.1 \times 0.1 = 0.65 \times 0.1 = \mathbf{0.065}$.

— or —

Multiplying a number by 0.1 once moves the number's decimal point one place to the left. So, multiplying a number by 0.1 *twice* moves the number's decimal point *two* places to the left.

So, $6.5 \times 0.1 \times 0.1 = \mathbf{0.065}$.

54. Since multiplication is associative and commutative, $0.1 \times 0.046 \times 0.1 = 0.046 \times 0.1 \times 0.1$. Multiplying a number by 0.1 twice moves the number's decimal point two places to the left.

So, $0.1 \times 0.046 \times 0.1 = 0.046 \times 0.1 \times 0.1 = \mathbf{0.00046}$.

55. Multiplying 831 by 0.1 three times moves the decimal point in 831 three places to the left.

So, $831 \times 0.1 \times 0.1 \times 0.1 = \mathbf{0.831}$.

56. Dividing by a number is the same as multiplying by that number's reciprocal. The reciprocal of 10 is $\frac{1}{10} = 0.1$, so dividing by 10 is the same as multiplying by 0.1.

So, $0.0345 \div 10 = 0.0345 \times 0.1 = \mathbf{0.00345}$.

57. Dividing by a number is the same as multiplying by that number's reciprocal. Since $0.1 = \frac{1}{10}$, the reciprocal of 0.1 is 10. So, dividing by 0.1 is the same as multiplying by 10.

So, $0.0067 \div 0.1 = 0.0067 \times 10 = \mathbf{0.067}$.

58. $(0.1)^2 = 0.1 \times 0.1$. To compute this product, we move the decimal point in 0.1 one place to the left.

So, $(0.1)^2 = \mathbf{0.01}$.

59. $(0.1)^3 = 0.1 \times 0.1 \times 0.1$. To compute this product, we move the decimal point in 0.1 two places to the left.

So, $(0.1)^3 = \mathbf{0.001}$.

60. We notice a pattern in the powers of 0.1.

$(0.1)^1 = 0.1$ has 1 digit right of the decimal point.
$(0.1)^2 = 0.01$ has 2 digits right of the decimal point.
$(0.1)^3 = 0.001$ has 3 digits right of the decimal point.

Since each copy of 0.1 shifts the decimal point one place to the left, we know this pattern continues. So, $(0.1)^6$ has 6 digits right of the decimal point: 5 zeros, followed by a 1.

Therefore, $(0.1)^6 = \mathbf{0.000001}$.

In general, $(0.1)^n$ has n digits after the decimal point: $(n-1)$ zeros, followed by a 1.

61. $2.3 \times (0.1)^2$ is the product of 2.3 and two copies of 0.1. So, we move the decimal point in 2.3 two places to the left. $2.3 \times (0.1)^2 = \mathbf{0.023}$.

62. Since $0.01 = 0.1 \times 0.1$, multiplying a number by 0.01 moves that number's decimal point two places to the left. So, $0.3 \times 0.01 = \mathbf{0.003}$.

63. $(0.1)^3 \times 404$ is the product of 404 and three copies of 0.1. So, we move the decimal point in 404 three places to the left: $(0.1)^3 \times 404 = \mathbf{0.404}$.

64. Since $0.001 = 0.1 \times 0.1 \times 0.1$, multiplying a number by 0.001 moves that number's decimal point three places to the left. So, $6.7 \times 0.001 = \mathbf{0.0067}$.

65. Multiplying a number by $(0.1)^2$ moves that number's decimal point two places to the left.

So, $(0.1)^2 \times 6.05 = \mathbf{0.0605}$.

66. Since $0.0001 = (0.1)^4$, multiplying a number by 0.0001 moves that number's decimal point four places to the left. So, $8,500 \times 0.0001 = 0.8500 = \mathbf{0.85}$.

67. Multiplying a number by $(0.1)^6$ moves that number's decimal point six places to the left.

So, $(0.1)^6 \times 4,000,000 = 4.000000 = \mathbf{4}$.

68. Since $0.00001 = (0.1)^5$, multiplying a number by 0.00001 moves that number's decimal point five places to the left. So, $75.25 \times 0.00001 = \mathbf{0.0007525}$.

69. To go from 1.2 to 0.012, we move the decimal point two places to the left. Multiplying a number by 0.01 moves the number's decimal point two places to the left.

So, $1.2 \times \mathbf{0.01} = 0.012$.

70. To go from 312.5 to 0.3125, we move the decimal point three places to the left. Multiplying a number by 0.001 moves the number's decimal point three places to the left.

So, $\mathbf{0.001} \times 312.5 = 0.3125$.

71. When we multiply the missing number by 0.001, the decimal point moves three places to the *left*, giving 0.03456. So, to find the missing number, we move the decimal point in 0.03456 three places to the *right*. This gives 34.56.

So, $0.001 \times \mathbf{34.56} = 0.03456$.

72. When we multiply the missing number by 0.0001, the decimal point moves four places to the *left*, giving 0.25. So, to find the missing number, we move the decimal point in 0.25 four places to the *right*. This gives 2,500.

So, $\mathbf{2,500} \times 0.0001 = 0.25$.

Multiplying Decimals 82-87

73. We write each number as a fraction, multiply, then convert back to decimal form:

$$0.2 \times 0.3 = \frac{2}{10} \times \frac{3}{10}$$
$$= \frac{6}{100}$$
$$= \mathbf{0.06}.$$

— *or* —

Writing 0.2 as 2×0.1 and 0.3 as 3×0.1, we have

$$0.2 \times 0.3 = (2 \times 0.1) \times (3 \times 0.1)$$
$$= (2 \times 3) \times (0.1 \times 0.1)$$
$$= 6 \times 0.01$$
$$= \mathbf{0.06}.$$

74. We have

$$0.4 \times 0.8 = \frac{4}{10} \times \frac{8}{10}$$
$$= \frac{32}{100}$$
$$= \mathbf{0.32}.$$

— *or* —

$$0.4 \times 0.8 = (4 \times 0.1) \times (8 \times 0.1)$$
$$= (4 \times 8) \times (0.1 \times 0.1)$$
$$= 32 \times 0.01$$
$$= \mathbf{0.32}.$$

75. We have

$$0.06 \times 0.11 = \frac{6}{100} \times \frac{11}{100}$$
$$= \frac{66}{10,000}$$
$$= \mathbf{0.0066}.$$

— *or* —

$$0.06 \times 0.11 = (6 \times 0.01) \times (11 \times 0.01)$$
$$= (6 \times 11) \times (0.01 \times 0.01)$$
$$= 66 \times 0.0001$$
$$= \mathbf{0.0066}.$$

76. We have

$$0.05 \times 0.3 = \frac{5}{100} \times \frac{3}{10}$$
$$= \frac{15}{1,000}$$
$$= \mathbf{0.015}.$$

— *or* —

$$0.05 \times 0.3 = (5 \times 0.01) \times (3 \times 0.1)$$
$$= (5 \times 3) \times (0.01 \times 0.1)$$
$$= 15 \times 0.001$$
$$= \mathbf{0.015}.$$

77. We have

$$0.006 \times 0.12 = \frac{6}{1,000} \times \frac{12}{100}$$
$$= \frac{72}{100,000}$$
$$= \mathbf{0.00072}.$$

— *or* —

$$0.006 \times 0.12 = (6 \times 0.001) \times (12 \times 0.01)$$
$$= (6 \times 12) \times (0.001 \times 0.01)$$
$$= 72 \times 0.00001$$
$$= \mathbf{0.00072}.$$

78. We have

$$0.07 \times 0.8 \times 0.002 = \frac{7}{100} \times \frac{8}{10} \times \frac{2}{1,000}$$
$$= \frac{112}{1,000,000}$$
$$= \mathbf{0.000112}.$$

— *or* —

$$0.07 \times 0.8 \times 0.002 = (7 \times 0.01) \times (8 \times 0.1) \times (2 \times 0.001)$$
$$= (7 \times 8 \times 2) \times (0.01 \times 0.1 \times 0.001)$$
$$= 112 \times 0.000001$$
$$= \mathbf{0.000112}.$$

79. Writing 0.0049 as a fraction, we have $\frac{49}{10,000}$. Since $49 = 7 \times 7$, and $10,000 = 100 \times 100$, we have

$$\begin{aligned} 0.0049 &= \frac{49}{10,000} \\ &= \frac{7 \times 7}{100 \times 100} \\ &= \frac{7}{100} \times \frac{7}{100} \\ &= \left(\frac{7}{100}\right)^2. \end{aligned}$$

So, the square of $\frac{7}{100}$ is 0.0049. As a decimal, $\frac{7}{100} = \mathbf{0.07}$.

— *or* —

We begin by writing 0.0049 as 49×0.0001. Since $49 = 7 \times 7$ and $0.0001 = 0.01 \times 0.01$, we have

$$\begin{aligned} 0.0049 &= 49 \times 0.0001 \\ &= (7 \times 7) \times (0.01 \times 0.01) \\ &= (7 \times 0.01) \times (7 \times 0.01) \\ &= 0.07 \times 0.07 \\ &= (0.07)^2. \end{aligned}$$

So, the square of **0.07** is 0.0049. As a fraction, $0.07 = \frac{7}{100}$.

80. The perimeter of the rectangle is 1 in, so $2(0.A + 0.B) = 1$. Therefore, $0.A + 0.B = \frac{1}{2} = 0.5$.

Since A and B are digits, A+B must be 5. This gives the following possibilities:

- A = 1, B = 4 (or A = 4, B = 1):
  In this case, the area of the rectangle is $0.1 \times 0.4 = 0.04$ square inches.

- A = 2, B = 3 (or A = 3, B = 2):
  In this case, the area of the rectangle is $0.2 \times 0.3 = 0.06$ square inches.

So, the largest possible area of the rectangle is **0.06** square inches.

81. 1.1 and 1.1 have a total of $1+1 = 2$ digits to the right of the decimal point.

So, we place the decimal point in 121 so that there are 2 digits to the right of the decimal point.

— *or* —

Since 1.1 is a little more than 1, we expect 1.1×1.1 to be a little more than $1 \times 1 = 1$. So, we place the decimal point in the only place that gives a number that is a little more than 1.

1.21

82. 10.1 and 1.001 have a total of $1+3 = 4$ digits to the right of the decimal point.

So, we place the decimal point in 101101 so that there are 4 digits to the right of the decimal point.

— *or* —

Since 10.1 is a little more than 10, and 1.001 is a little more than 1, we expect 10.1×1.001 to be a little more than $10 \times 1 = 10$. So, we place the decimal point in the only place that gives a number that is a little more than 10.

10.1101

83. 33.34 and 333.4 have a total of $2+1 = 3$ digits to the right of the decimal point.

So, we place the decimal point in 11115556 so that there are 3 digits to the right of the decimal point.

11115.556

84. 1,428.57 and 0.007 have a total of $2+3 = 5$ digits to the right of the decimal point.

So, we place the decimal point in 999999 so that there are 5 digits to the right of the decimal point.

9.99999

85. 0.03 and 0.8 have a total of $2+1 = 3$ digits to the right of the decimal point.

So, we move the decimal point in $3 \times 8 = 24$ so that there are 3 digits to the right of the decimal point.

So, $0.03 \times 0.8 = \mathbf{0.024}$.

86. 0.012 and 9 have a total of $3+0 = 3$ digits to the right of the decimal point.

So, we move the decimal point in $12 \times 9 = 108$ so that there are 3 digits to the right of the decimal point.

So, $0.012 \times 9 = \mathbf{0.108}$.

87. 0.002 and 0.0006 have a total of $3+4 = 7$ digits to the right of the decimal point.

So, we move the decimal point in $2 \times 6 = 12$ so that there are 7 digits to the right of the decimal point.

So, $0.002 \times 0.0006 = \mathbf{0.0000012}$.

88. 0.0404 and 0.07 have a total of $4+2 = 6$ digits to the right of the decimal point.

So, we move the decimal point in $404 \times 7 = 2,828$ so that there are 6 digits to the right of the decimal point.

So, $0.0404 \times 0.07 = \mathbf{0.002828}$.

89. 0.2 and 0.5 have a total of $1+1 = 2$ digits to the right of the decimal point.

So, we move the decimal point in $2 \times 5 = 10$ so that there are 2 digits to the right of the decimal point, including the trailing zero. This gives 0.10.

After we have placed the decimal point, we can remove the trailing zero. So, $0.2 \times 0.5 = 0.10 = \mathbf{0.1}$.

90. 0.06 and 0.25 have a total of $2+2=4$ digits to the right of the decimal point.

$$0.\underbrace{06}_{2} \times 0.\underbrace{25}_{2}$$

So, we move the decimal point in $6\times25=150$ so that there are 4 digits to the right of the decimal point, including the trailing zero. This gives 0.0150.

$$0.\underbrace{0150}_{4}$$

After we have placed the decimal point, we can remove the trailing zero. So, $0.06\times0.25=0.0150=\textbf{0.015}$.

91. 0.075 and 0.8 have a total of $3+1=4$ digits to the right of the decimal point.

So, we move the decimal point in $75\times8=600$ so that there are 4 digits to the right of the decimal point, including the trailing zeros. This gives 0.0600.

After placing the decimal point, we can remove the trailing zeros: $0.075\times0.8=0.0600=\textbf{0.06}$.

92. 0.00125 and 0.032 have a total of $5+3=8$ digits to the right of the decimal point.

So, we move the decimal point in $125\times32=4,000$ so that there are 8 digits to the right of the decimal point, including the trailing zeros. This gives 0.00004000.

After placing the decimal point, we can remove the trailing zeros: $0.00125\times0.032=0.00004000=\textbf{0.00004}$.

93. The numbers in $0.9\times0.8\times0.7\times0.6\times0.5\times0.4\times0.3\times0.2\times0.1$ have a total of 9 digits to right of the decimal point. So, there are 9 digits to the right of the decimal point in the product. However, some of those digits may be trailing zeros, which we do not count.

The number of trailing zeros is equal to the number of zeros at the end of $9!=9\times8\times7\times6\times5\times4\times3\times2\times1$.

The number of zeros at the end of 9! is given by the largest power of 10 that is a factor of (9!). We can pair a 2 and a 5 in $9\times8\times7\times6\times5\times4\times3\times2\times1$ to make a 10. Since there are no more 5's, we cannot make any more 10's. So, 10^1 is the largest power of 10 that is a factor of (9!). Therefore, there is 1 zero at the end of (9!).

So, the product $0.9\times0.8\times0.7\times0.6\times0.5\times0.4\times0.3\times0.2\times0.1$ has $9-1=\textbf{8}$ digits to the right of the decimal point.

In fact, 9! = 362,880, and
0.9×0.8×0.7×0.6×0.5×0.4×0.2×0.1 = 0.00036288.

94. 0.3 has 1 digit to the right of the decimal point, and 0.07 has 2 digits to the right of the decimal point.

So, multiplying 15 copies of 0.3 and 15 copies of 0.07 gives $(15\times1)+(15\times2)=15+30=45$ digits to the right of the decimal point, including any trailing zeros.

Since neither 3 nor 7 has factors of 2 or 5, no product of 3's and 7's will have trailing zeros. So, there are no trailing zeros to remove after the decimal point.

Therefore, $(0.3)^{15}\times(0.07)^{15}$ has **45** digits to the right of the decimal point.

95. 0.6 has 1 digit to the right of the decimal point, and 0.05 has 2 digits to the right of the decimal point.

So, multiplying 15 copies of 0.6 and 15 copies of 0.05

gives $(15\times1)+(15\times2)=15+30=45$ digits to the right of the decimal point, including trailing zeros.

To count the number of trailing zeros that are removed, we count the zeros at the end of $6^{15}\times5^{15}$. The product of 15 copies of 6 and 15 copies of 5 is the same as the product of 15 copies of 6×5. Since $6\times5=30$ has 1 zero, the product of 15 copies of 6×5 has 15 zeros.

Therefore, $(0.6)^{15}\times(0.05)^{15}$ has $45-15=\textbf{30}$ digits to the right of the decimal point.

— *or* —

$(0.6)^{15}\times(0.05)^{15}$ is the product of 15 copies of 0.6 and 15 copies of 0.05. We can pair each 0.6 with a 0.05 to get the product of 15 copies of $0.6\times0.05=0.030=0.03$.

So, $(0.6)^{15}\times(0.05)^{15}=(0.03)^{15}$.

Since 0.03 has 2 digits to the right of the decimal point with no trailing zeros, the product of 15 copies of 0.03 has $15\times2=30$ digits to the right of the decimal point with no trailing zeros.

Therefore, $(0.6)^{15}\times(0.05)^{15}=(0.03)^{15}$ has **30** digits to the right of the decimal point.

96. Since 0.7 and 0.□ have a total of 2 digits to the right of the decimal point, and their product 0.5□ also has 2 digits to the right of the decimal point, we can ignore the decimal points and consider the following equation:

$$7\times\boxed{}=5\boxed{}.$$

8 is the only number we can multiply by 7 to get a product in the 50's.

Since $7\times8=56$, we have

$$0.7\times0.\boxed{8}=0.5\boxed{6}.$$

97. Since 10.□ and 0.□ have a total of 2 digits to the right of the decimal point, and their product 3.□6 also has 2 digits to the right of the decimal point, we can ignore the decimal points and consider the following equation:

$$10\boxed{}\times\boxed{}=3\boxed{}6.$$

3 is the only digit we can multiply 10□ by to get a product in the 300's. So, we have $10\boxed{}\times\boxed{3}=3\boxed{}6$.

Then, the missing units digit in 10□ must be 2 in order to get a units digit of 6 in 3□6. This gives $10\boxed{2}\times\boxed{3}=3\boxed{0}6$.

Therefore, we have

$$10.\boxed{2}\times0.\boxed{3}=3.\boxed{0}6.$$

98. All together, 0.□×0.2 has 1 more digit to the right of the decimal point than the product 0.□. To make the number of digits to the right of the decimal point on both sides of the equation match, we can write the equation as 0.□×0.2=0.□0.

Then, we can temporarily ignore the decimal points and consider the equation □×2=□0.

5 is the only digit that can fill the blank in □×2 to give a product that ends in zero. Since $\boxed{5}\times2=\boxed{1}0$, we have

$$0.5\times0.2=0.10=0.1.$$

We fill in the blanks as shown:

$$0.\boxed{5}\times0.2=0.\boxed{1}.$$

99. All together, $0.6 \times 2.\square$ has 1 more digit to the right of the decimal point than the product $\square.\square$. To make the number of digits to the right of the decimal point on both sides of the equation match, we can write the equation as $0.6 \times 2.\square = \square.\square 0$.

Then, we can temporarily ignore the decimal points and consider the equation $6 \times 2\square = \square\square 0$.

For the product of 6 and another integer to end in zero, the integer must be a multiple of 5, and therefore ends in 0 or 5. However, numbers in this problem cannot have trailing zeros, so the integer we multiply by 6 must end in 5. Since $6 \times 2\boxed{5} = \boxed{1}\boxed{5}0$, we have

$$0.6 \times 2.5 = 1.50 = 1.5.$$

We fill in the blanks as shown.

$$0.6 \times 2.\boxed{5} = \boxed{1}.\boxed{5}.$$

100. All together, $0.\square \times 4.\square$ has the same number of digits to the right of the decimal point as the product $0.\square 3$. So, we can temporarily ignore the decimal points and consider the following equation:

$$\square \times 4\square = \square 3.$$

If the digit in the first blank is 3 or greater, then the product will be $3 \times 40 = 120$ or greater. However, the product is a two-digit number. So, the digit in the first blank is 1 or 2.

If the digit in the first blank is 2, then the product will be even. However, $\square 3$ has units digit 3, so it is odd. Therefore, the digit in the first blank is 1.

$$\boxed{1} \times 4\square = \square 3.$$

Then, the digit in the second blank must be 3, and the digit in the final blank must be 4.

$$\boxed{1} \times 4\boxed{3} = \boxed{4}3.$$

We fill in the blanks as shown.

$$0.\boxed{1} \times 4.\boxed{3} = 0.\boxed{4}3.$$

101. All together, $0.0\square \times 0.0\square$ has 1 more digit to the right of the decimal point than the product 0.003. To make the number of digits to the right of the decimal point on both sides of the equation match, we can write the equation as $0.0\square \times 0.0\square = 0.0030$.

Then, we can temporarily ignore the decimal points and consider the equation $\square \times \square = 30$. There is only one pair of digits with product 30: $\boxed{5} \times \boxed{6} = 30$.

So, we have

$$0.05 \times 0.06 = 0.0030 = 0.003.$$

We fill in the blanks as shown.

$$0.0\boxed{5} \times 0.0\boxed{6} = 0.003.$$

Since multiplication is commutative, you may have instead written $0.0\boxed{6} \times 0.0\boxed{5} = 0.003$.

102. All together, $0.\square 5 \times 0.0\square$ has 2 more digits to the right of the decimal point than the product 0.03. To make the number of digits to the right of the decimal point on both sides of the equation match, we can write the equation as $0.\square 5 \times 0.0\square = 0.0300$.

Then, we can temporarily ignore the decimal points and consider the equation $\square 5 \times \square = 300$.

The only factor pairs of 300 that include a number with units digit 5 are 5×60, 15×20, 25×12, and 75×4. Of these, only 75×4 can be used to fill in the missing digits.

Since $\boxed{7}5 \times \boxed{4} = 300$, we have

$$0.75 \times 0.04 = 0.0300 = 0.03.$$

We fill in the blanks as shown.

$$0.\boxed{7}5 \times 0.0\boxed{4} = 0.03.$$

103. All together, $0.3\square \times 0.0\square$ has 1 more digit to the right of the decimal point than the product 0.007. To make the number of digits to the right of the decimal point on both sides of the equation match, we can write the equation as $0.3\square \times 0.0\square = 0.0070$.

Then, we can temporarily ignore the decimal points and consider the equation $3\square \times \square = 70$.

The only factor pair of 70 that includes a number with tens digit 3 is 35×2. Since $\boxed{3}5 \times \boxed{2} = 70$, we have

$$0.35 \times 0.02 = 0.0070 = 0.007.$$

We fill in the blanks as shown.

$$0.3\boxed{5} \times 0.0\boxed{2} = 0.007.$$

104. All together, $0.\square\square\square \times 0.\square$ has 3 more digits to the right of the decimal point than the product 0.7. To make the number of digits to the right of the decimal point on both sides of the equation match, we can write the equation as $0.\square\square\square \times 0.\square = 0.7000$.

Then, we can temporarily ignore the decimal points and consider the equation $\square\square\square \times \square = 7,000$. Since $1,000 \times 7 = 7,000$, the one-digit number in the equation $\square\square\square \times \square = 7,000$ must be greater than 7 in order for the other missing number to have fewer than four digits. So, the one-digit number is either 8 or 9. Since 9 is not a factor of 7,000, the one-digit number must be 8.

So, we have $\square\square\square \times 8 = 7,000$. Therefore, the missing three-digit number is $7,000 \div 8 = 875$.

Since $\boxed{8}\boxed{7}\boxed{5} \times \boxed{8} = 7,000$, we have

$$0.875 \times 0.8 = 0.7000 = 0.7.$$

We fill in the blanks as shown.

$$0.\boxed{8}\boxed{7}\boxed{5} \times 0.\boxed{8} = 0.7.$$

— *or* —

We consider the equation $\square\square\square \times \square = 7,000$. The prime factorization of 7,000 is $2^3 \times 5^3 \times 7$. Since numbers in this problem cannot have trailing zeros, neither of the missing numbers in the product $\square\square\square \times \square$ can have a factor of 2 *and* a factor of 5. This gives the following possibilities:

$$(2^3 \times 7) \times (5^3) \quad \text{or} \quad (5^3 \times 7) \times (2^3).$$

Only the product $(5^3 \times 7) \times (2^3)$ includes a one-digit number: $2^3 = 8$.

So, we have $(5^3 \times 7) \times (2^3) = 875 \times 8 = 7,000$. Therefore,

$$0.875 \times 0.8 = 0.7000 = 0.7.$$

We fill in the blanks as shown.

$$0.\boxed{8}\boxed{7}\boxed{5} \times 0.\boxed{8} = 0.7.$$

105. All together, $1.\square \times 0.\square\square\square$ has 2 more digits to the right of the decimal point than the product 0.04. To make the number of digits to the right of the decimal point on both sides of the equation match, we can write the equation as $1.\square \times 0.\square\square\square = 0.0400$.

Then, we can temporarily ignore the decimal points and consider the equation $1\square \times \square\square\square = 400$. The number $1\square$ is at least 10 and at most 19, and is also a factor of 400. The only numbers from 10 to 19 that are factors of 400 are 10 and 16. However, $1\square$ cannot be 10 since numbers in this problem cannot have trailing zeros.

So, we have $1\boxed{6} \times \square\square\square = 400$. Therefore, the other missing number is $400 \div 16 = 25$. Since 25 has only two digits, we write a leading 0 in the hundreds place.

Since $1\boxed{6} \times \boxed{0}\boxed{2}\boxed{5} = 400$, we have

$$1.6 \times 0.025 = 0.0400 = 0.04.$$

We fill in the blanks as shown.

$$1.\boxed{6} \times 0.\boxed{0}\boxed{2}\boxed{5} = 0.04.$$

106. Since $200 = 2 \times 100$, multiplying by 200 is the same as multiplying by 2, then shifting the decimal point in the product 2 places to the *right*.

Similarly, since $0.009 = 9 \times 0.001$, multiplying by 0.009 is the same as multiplying by 9, then shifting the decimal point in the product 3 places to the *left*.

So, to compute 200×0.009, we can multiply $2 \times 9 = 18$, then shift the decimal point 2 places to the right and 3 places to the left. This is the same as shifting the decimal point 1 place to the left:

$$1.\underset{\curvearrowleft}{8}$$

So, $200 \times 0.009 = \mathbf{1.8}$.

107. Since $0.04 = 4 \times 0.01$, multiplying by 0.04 is the same as multiplying by 4, then shifting the decimal point in the product 2 places to the *left*.

Similarly, since $13,000 = 13 \times 1,000$, multiplying by 13,000 is the same as multiplying by 13, then shifting the decimal point in the product 3 places to the *right*.

So, to compute $0.04 \times 13,000$, we can multiply $4 \times 13 = 52$, then shift the decimal point 2 places to the left and 3 places to the right. This is the same as shifting the decimal point 1 place to the right:

$$52\underset{\curvearrowright}{0}.$$

So, $0.04 \times 13,000 = \mathbf{520}$.

108. Multiplying by 0.00025 is the same as multiplying by 25, then shifting the decimal point 5 places to the left.

Multiplying by 800 is the same as multiplying by 8, then shifting the decimal point 2 places to the right.

So, to compute 0.00025×800, we multiply $25 \times 8 = 200$, then shift the decimal point 3 places to the left.

So, $0.00025 \times 800 = \mathbf{0.2}$.

109. We begin by computing $3 \times 3 \times 3 = 27$. Then we determine where to place the decimal point.

Multiplying by 300 shifts the decimal point right 2 places.
Multiplying by 0.03 shifts the decimal point left 2 places.
Multiplying by 0.003 shifts the decimal point left 3 places.

All together, we shift the decimal point in 27 left 3 places. So, $300 \times 0.03 \times 0.003 = \mathbf{0.027}$.

110. We begin by computing $4 \times 12 \times 1 = 48$. Then we determine where to place the decimal point.

Multiplying by 40 shifts the decimal point right 1 place.
Multiplying by 0.012 shifts the decimal point left 3 places.
Multiplying by 1,000 shifts the decimal point right 3 places.

All together, we shift the decimal point in 48 right 1 place. So, $40 \times 0.012 \times 1,000 = \mathbf{480}$.

— *or* —

We see that $0.012 \times 1,000 = 12$.
So, $40 \times 0.012 \times 1,000 = 40 \times 12 = \mathbf{480}$.

111. We begin by computing $25 \times 4 \times 2 = 200$. Then we determine where to place the decimal point.

Multiplying by 0.025 shifts the decimal point left 3 places.
Multiplying by 0.04 shifts the decimal point left 2 places.
Multiplying by 2,000 shifts the decimal point right 3 places.

All together, we shift the decimal point in 200 left 2 places. So, $0.025 \times 0.04 \times 2,000 = \mathbf{2}$.

— *or* —

Since multiplication is commutative and associative,

$$0.025 \times 0.04 \times 2,000 = 0.04 \times (0.025 \times 2,000).$$

We see that $0.025 \times 2,000 = 50$.
So, $0.04 \times (0.025 \times 2,000) = 0.04 \times 50 = \mathbf{2}$.

112. $(300)^{10} \times (0.002)^{10}$ is the product of 10 copies of 300 and 10 copies of 0.002.

Multiplying by 300 shifts the decimal point right 2 places. So, multiplying by 10 copies of 300 shifts the decimal point right $2 \times 10 = 20$ places.

Multiplying by 0.002 shifts the decimal point left 3 places. So, multiplying by 10 copies of 0.002 shifts the decimal point left $3 \times 10 = 30$ places.

All together, we shift the decimal point in the product left 10 places. So, to compute $(300)^{10} \times (0.002)^{10}$, we multiply $3^{10} \times 2^{10}$, which equals some integer that does not end in zero. Then, we shift the decimal point 10 places to the left, which gives a product with **10** digits to the right of the decimal point.

In fact, $(300)^{10} \times (0.002)^{10} = 0.0060466176$.

— *or* —

$(300)^{10} \times (0.002)^{10}$ is the product of 10 copies of 300 and 10 copies of 0.002. We can pair each 300 with a 0.002 to get the product of 10 copies of $300 \times 0.002 = 0.6$. So, $(300)^{10} \times (0.002)^{10} = (0.6)^{10}$.

Each time we multiply by 0.6, the decimal point in the product shifts 1 place to the left. So, the product of 10 copies of 0.6 shifts the decimal point in the product $1 \times 10 = 10$ places to the left.

So, to compute $(300)^{10} \times (0.002)^{10} = (0.6)^{10}$, we compute 6^{10}, which equals some integer that does not end in zero. Then, we shift the decimal point left 10 places, which gives a product with **10** digits to the right of the decimal point.

In fact, $(0.6)^{10} = 0.0060466176$.

113. $(400)^{10} \times (0.0025)^{10}$ is the product of 10 copies of 400 and 10 copies of 0.0025.

Multiplying by 400 shifts the decimal point right 2 places. So, multiplying by 10 copies of 400 shifts the decimal point right $2 \times 10 = 20$ places.

Multiplying by 0.0025 shifts the decimal point left 4 places. So, multiplying by 10 copies of 0.0025 shifts the decimal point left $4 \times 10 = 40$ places.

All together, we shift the decimal point in the product left 20 places. So, to compute $(400)^{10} \times (0.0025)^{10}$, we multiply $4^{10} \times 25^{10}$, then shift the decimal point 20 places to the left.

Since $4 \times 25 = 100$, we can pair ten 4's with ten 25's to get $4^{10} \times 25^{10} = (4 \times 25)^{10} = 100^{10}$. Each copy of 100 in 100^{10} adds two trailing zeros to the product. So, 100^{10} equals 1 followed by $2 \times 10 = 20$ trailing zeros:

$$100,000,000,000,000,000,000.$$

Then, shifting the decimal point left 20 places gives us 1. So, $(400)^{10} \times (0.0025)^{10} = \mathbf{1}$.

— *or* —

$(400)^{10} \times (0.0025)^{10}$ is the product of 10 copies of 400 and 10 copies of 0.0025. We can pair each 400 with a 0.0025 to get the product of 10 copies of 400×0.0025. So, $(400)^{10} \times (0.0025)^{10} = (400 \times 0.0025)^{10}$.

Since $400 \times 0.0025 = 1$, we have $(400 \times 0.0025)^{10} = 1^{10} = \mathbf{1}$.

114. To compute the height of a stack of 40 bricks in which each brick is 0.64 centimeters tall, we multiply 40×0.64.

Multiplying by 40 shifts the decimal point right 1 place. Multiplying by 0.64 shifts the decimal point left 2 places.

So, to compute 40×0.64, we multiply $4 \times 64 = 256$, then shift the decimal point a total of 1 place to the left.

So, the height of the 40-brick stack is $40 \times 0.64 = \mathbf{25.6}$ cm.

Check: Since 0.64 is a little more than half, we expect 40×0.64 to be a little more than half of 40. ✓

115. The volume of a cube with 0.4-cm edges is $0.4 \times 0.4 \times 0.4$. To compute this product, we multiply $4 \times 4 \times 4 = 64$, then move the decimal point 3 places to the left.

Therefore, the volume of the cube is $0.4 \times 0.4 \times 0.4 = \mathbf{0.064}$ cubic cm.

116. Since 1 inch is 2.54 centimeters, 0.3 inches is 0.3×2.54 centimeters.

Since $3 \times 254 = 762$, we have $0.3 \times 2.54 = 0.762$. So, a 0.3-inch tall gentlebug is **0.762** centimeters tall.

117. After working for 6 hours, Timmy will earn 12.25×6 dollars. Since $1,225 \times 6 = 7,350$, we have $12.25 \times 6 = 73.5$.

So, Timmy will earn **$73.50**.

— *or* —

After working for 6 hours, Timmy will earn 12.25×6 dollars. We use the distributive property:

$$
\begin{aligned}
12.25 \times 6 &= (12 + 0.25) \times 6 \\
&= (12 \times 6) + (0.25 \times 6) \\
&= 72 + 1.5 \\
&= 73.5.
\end{aligned}
$$

So, Timmy will earn **$73.50**.

118. We consider the computation using fractions.
$0.125 = \frac{125}{1,000} = \frac{1}{8}$, and $0.01 = \frac{1}{100}$. So, we have $\frac{1}{8} \times a = \frac{1}{100}$. Therefore, $a = \frac{1}{100} \div \frac{1}{8}$.

Dividing by a number is the same as multiplying by that number's reciprocal. The reciprocal of $\frac{1}{8}$ is 8. Therefore, $a = \frac{1}{100} \times 8 = \frac{8}{100} = \mathbf{0.08}$.
As a fraction, we write $a = \frac{8}{100} = \frac{2}{25}$.

— *or* —

We know that $125 \times 8 = 1,000$. Since 0.125 has 3 digits to the right of the decimal point, $0.125 \times 8 = 1$.

However, the product we seek is 0.01. To get 0.01, we move the decimal point in 8 two places to the left. This gives $0.125 \times 0.08 = 0.01$.

So, $a = \mathbf{0.08}$.

As a fraction, we write $a = \frac{8}{100} = \frac{2}{25}$.

119. Rather than computing the areas of the two rectangles separately, then adding the results, we notice that each rectangle has a side with length 2.1 m. We can join the two rectangles along this side to create a larger rectangle with the same area as both smaller rectangles combined.

This larger rectangle has height 2.1 m and width $4.473 + 3.527 = 8$ m. So, its area is $2.1 \times 8 = \mathbf{16.8}$ sq m.

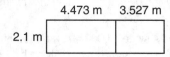
Pyramid Descent 88-89

120. Ignoring decimal points and trailing zeros, the top block contains a 1, each block in the middle row contains a 2, and each block in the bottom row contains a 3. So, every path has a product of $1 \times 2 \times 3 = 6$.

The product we seek is 0.06. To go from 6 to 0.06, we move the decimal point two places to the left. The direction in which we move the decimal point and the number of times we move it is determined by the number of trailing zeros and the number of digits after the decimal point in each path.

We consider the placement of the decimal point for each path, then circle the numbers in the correct path as shown.

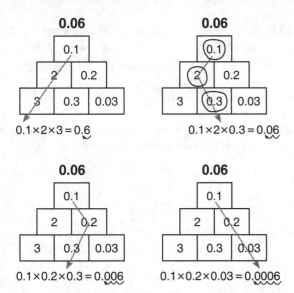

0.1×2×3 = 0.6

0.1×2×0.3 = 0.06

0.1×0.2×0.3 = 0.006

0.1×0.2×0.03 = 0.0006

121. Ignoring decimal points and trailing zeros, each block in the pyramid contains a 2. So, every path has a product of 2×2×2 = 8.

The product we seek is 0.008. To go from 8 to 0.008, we move the decimal point 3 places to the left. The path below is the only path in which the decimal point moves a total of 3 places left.

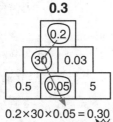

0.2×0.02×2 = 0.008

122. Ignoring decimal points and trailing zeros, every path in this pyramid has a product of 2×3×5 = 30. To go from 30 to 0.3, we move the decimal point 2 places to the left. The path below is the only path in which the decimal point moves a total of 2 places left.

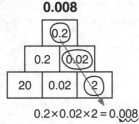

0.2×30×0.05 = 0.30

123. Ignoring decimal points and trailing zeros, every path in this pyramid has a product of 3×3×3×3 = 81. To go from 81 to 8.1, we move the decimal point 1 place to the left. The path below is the only path in which the decimal point moves a total of 1 place left.

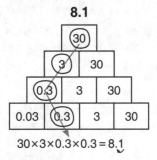

30×3×0.3×0.3 = 8.1

124. Ignoring decimal points and trailing zeros, every path in this pyramid has a product of 1×7×11×13 = 1,001. To go from 1,001 to 0.1001, we move the decimal point 4 places to the left. The path below is the only path in which the decimal point moves a total of 4 places left.

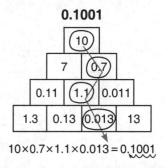

10×0.7×1.1×0.013 = 0.1001

125. Ignoring decimal points and trailing zeros, every path in this pyramid has a product of 2×5×2×5 = 100. To go from 100 to 0.01, we move the decimal point 4 places to the left. The path below is the only path in which the decimal point moves a total of 4 places left.

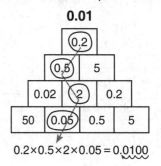

0.2×0.5×2×0.05 = 0.0100

126. Ignoring decimal points and trailing zeros, every path in this pyramid is a product of 2's and 3's, so the path must have a product of 24. To go from 24 to 0.024, we move the decimal point 3 places to the left. The path below is the only path in which the decimal point moves a total of 3 places left.

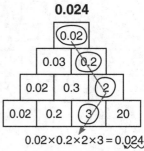

0.02×0.2×2×3 = 0.024

127. Ignoring decimal points and trailing zeros, every block in this pyramid contains a 1, a 3, or a 4. So, to get a product of 12, we need exactly one 3 and exactly one 4. There are only four paths with one 3 and one 4, as shown.

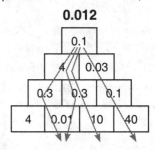

Then, to go from 12 to 0.012, we move the decimal point 3 places to the left. Among the four paths, the only one in which the decimal point moves a total of 3 places left is shown below.

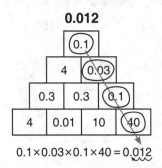

0.012

$$0.1 \times 0.03 \times 0.1 \times 40 = 0.012$$

128. Ignoring decimal points and trailing zeros, every block in the pyramid contains a 1, a 24, or a 25. The only nonzero digit in the product we seek is 6. We can only get a product whose only nonzero digit is 6 by multiplying $24 \times 25 = 600$. So, we look for paths with exactly one 24 and exactly one 25. There are four such paths:

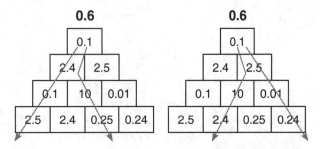

0.6   **0.6**

Then, to go from 600 to 0.6, we move the decimal point 3 places to the left. Among the paths shown above, the only one in which the decimal point moves a total of 3 places left is shown below.

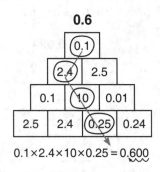

0.6

$$0.1 \times 2.4 \times 10 \times 0.25 = 0.600$$

DECIMALS

Estimation   90-91

For the problems below, you may have used different estimates than the ones given to arrive at the same final answers.

129. Since 0.0198 is about 0.02, and 407.6 is about 400, we estimate that 0.0198×407.6 is about $0.02 \times 400 = 8$.

So, 0.0198×407.6 is closer to **8** than to 80.

130. Since 523.9 is about 500, and 0.0031 is about 0.003, we estimate that 523.9×0.0031 is about $500 \times 0.003 = 1.5$.

So, 523.9×0.0031 is closer to **1.5** than to 15.

131. Since 0.068 is about 0.07 and 297.4 is about 300, we estimate that 0.068×297.4 is about $0.07 \times 300 = 21$.

Among our choices, only 20.2232 is close to 21.

2.2232   (20.2232)   30.2232   180.2232   202.232

132. 0.9594 is close to 1. So, we estimate the value of each answer choice and circle the one that is close to 1.

- $2.46 \times 0.039 \approx 2.5 \times 0.04 = 0.1$.
- $0.82 \times 11.7 \approx 1 \times 10 = 10$.
- $0.018 \times 5.33 \approx 0.02 \times 5 = 0.1$.
- $0.234 \times 4.1 \approx 0.25 \times 4 = 1$.

Among these choices, only 0.234×4.1 is close to 1.

2.46×0.039 0.82×11.7 0.018×5.33 (0.234×4.1)

133. We estimate each product.

- $50.03 \times 0.0819 \approx 50 \times 0.08 = 4$.
- $0.68 \times 0.392 \approx 0.7 \times 0.4 = 0.28$.
- $0.01207 \times 42.9 \approx 0.012 \times 40 = 0.48$.
- $0.054 \times 102.29 \approx 0.05 \times 100 = 5$.

We use these estimates to connect each product with its location on the number line, as shown below.

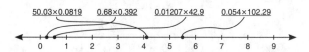

134. 0.9914 is less than 1, so 0.9914×0.52 is less than $1 \times 0.52 = 0.52$.

$$0.9914 \times 0.52 \;(<)\; 0.52$$

135. 3.911 is less than 4, and 0.049 is less than 0.05. So, 3.911×0.049 is less than $4 \times 0.05 = 0.2$.

$$3.911 \times 0.049 \;(<)\; 0.2$$

136. 9.008 is greater than 9, and 0.7134 is greater than 0.7. So, 9.008×0.7134 is greater than $9 \times 0.7 = 6.3$.

$$9.008 \times 0.7134 \;(>)\; 6.3$$

137. 98.05 is less than 100, so 98.05×0.00852 is less than $100 \times 0.00852 = 0.852$.

In other words, 0.852 is greater than 98.05×0.00852.

$$0.852 \;(>)\; 98.5 \times 0.00852$$

138. 0.029 is less than 0.03, and 0.38 is less than 0.4. So, 0.029×0.38 is less than $0.03 \times 0.4 = 0.012$.

2.05 is greater than 2, and 0.0061 is greater than 0.006. So, 2.05×0.0061 is greater than $2 \times 0.006 = 0.012$.

Since $0.029 \times 0.38 < 0.012$, and $2.05 \times 0.0061 > 0.012$, we have

$$0.029 \times 0.38 \;(<)\; 2.05 \times 0.0061.$$

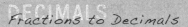
139. 3.208 is greater than 3, and 0.768 is greater than 0.7.
So, 3.208×0.768 is greater than 3×0.7 = 2.1.

4.95 is less than 5, and 0.3809 is less than 0.4.
So, 4.95×0.3809 is less than 5×0.4 = 2.

Since 3.208×0.768>2.1, and 4.95×0.3809<2, we have

$$3.208×0.768 \; \text{(>)} \; 4.95×0.3809.$$

140. 0.0625 is greater than 0.06, and 48 is greater than 40.
So, 0.0625×48 is greater than 0.06×40 = 2.4.

0.0625 is less than 0.07, and 48 is less than 50.
So, 0.0625×48 is less than 0.07×50 = 3.5.

So, 0.0625×48>2.4 and 0.0625×48<3.5. Since the problem tells us that 0.0625×48 is equal to an integer, that integer must be **3**.

— *or* —

0.0625 is about 0.06, and 48 is about 50. So, we estimate that 0.0625×48 is about 0.06×50 = 3. Since the problem tells us that 0.0625×48 is equal to an integer, that integer must be **3**.

141. 4.104 is greater than 4, and 5.724 is greater than 5.
So, 4.104×5.724 is greater than 4×5 = 20.

Also, 4.104 is less than 5, and 5.724 is less than 6.
So, 4.104×5.724 is less than 5×6 = 30.

Therefore, 4.104×5.724 is between 20 and 30. So, there are **7** multiples of 10 between 4.104×5.724 and 99:

30, 40, 50, 60, 70, 80, and 90.

142. We estimate each product.

- 0.028×7.03 ≈ 0.03×7 = 0.21.
- 0.982×0.41 ≈ 1×0.4 = 0.4.
- 0.62×1.571 ≈ 0.6×1.5 = 0.9.
- 0.76×0.259 ≈ 0.8×0.25 = 0.2.

Among these estimates, only 0.028×7.03 ≈ 0.21 and 0.76×0.259 ≈ 0.2 are close. So, we circle these two products as shown below.

 (0.028×7.03)   0.982×0.41   0.62×1.571   (0.76×0.259)

In fact, 0.028×7.03 = 0.76×0.259 = 0.19684.

143. Since 0.0058 is close to 0.006, and 21.3 is close to 20, Ralph's answer should be close to 0.006×20 = 0.12.

The answer Ralph got was 1.2354, which is about 10 times bigger than our estimate of 0.12. So, Ralph's answer is not reasonable.

In fact, 0.0058×21.3 = 0.12354. So, Ralph's answer was off by an entire place value. This could have happened by placing a decimal point in the wrong spot, or by forgetting one of the 0's after the decimal point in 0.0058. These are easy mistakes to make! This is why it's always a good idea to check the reasonableness of your answer, even when using a calculator!

144. Multiplying the numerator and denominator of $\frac{13}{50}$ by 2 gives an equivalent fraction with denominator 100.

$$\frac{13}{50} \overset{×2}{\underset{×2}{=}} \frac{26}{100}$$

So, $\frac{13}{50} = \frac{26}{100} = \mathbf{0.26}$.

145. Multiplying the numerator and denominator of $\frac{3}{4}$ by 25 gives an equivalent fraction with denominator 100.

$$\frac{3}{4} \overset{×25}{\underset{×25}{=}} \frac{75}{100}$$

So, $\frac{3}{4} = \frac{75}{100} = \mathbf{0.75}$.

146. Multiplying the numerator and denominator of $\frac{8}{5}$ by 2 gives an equivalent fraction with denominator 10.

$$\frac{8}{5} \overset{×2}{\underset{×2}{=}} \frac{16}{10}$$

So, $\frac{8}{5} = \frac{16}{10} = 1\frac{6}{10} = \mathbf{1.6}$.

— *or* —

We write $\frac{8}{5}$ as the mixed number $1\frac{3}{5}$. Multiplying the numerator and denominator of $\frac{3}{5}$ by 2 gives an equivalent fraction with denominator 10.

$$\frac{3}{5} \overset{×2}{\underset{×2}{=}} \frac{6}{10}$$

So, $\frac{8}{5} = 1\frac{3}{5} = 1\frac{6}{10} = \mathbf{1.6}$.

147. Multiplying the numerator and denominator of $\frac{7}{25}$ by 4, we have $\frac{7}{25} = \frac{28}{100} = \mathbf{0.28}$.

148. Multiplying the numerator and denominator of $\frac{51}{20}$ by 5, we have $\frac{51}{20} = \frac{255}{100} = 2\frac{55}{100} = \mathbf{2.55}$.

149. Neither 10 nor 100 are divisible by 8. However, 1,000÷8 = 125. So, multiplying the numerator and denominator of $\frac{5}{8}$ by 125, we have $\frac{5}{8} = \frac{625}{1,000} = \mathbf{0.625}$.

150. Multiplying the numerator and denominator of $\frac{9}{125}$ by 8, we have $\frac{9}{125} = \frac{72}{1,000} = \mathbf{0.072}$.

151. Multiplying the numerator and denominator of $\frac{3}{40}$ by 25, we have $\frac{3}{40} = \frac{75}{1,000} = \mathbf{0.075}$.

152. We write $\frac{33}{8}$ as the mixed number $4\frac{1}{8}$. Multiplying the numerator and denominator of $\frac{1}{8}$ by 125, we have $4\frac{1}{8} = 4\frac{125}{1,000} = \mathbf{4.125}$.

153. We write $\frac{13}{2^{99}\times 5^{101}}$ as an equivalent fraction with a denominator that is a power of 10.

Since $2^{99}\times 5^{101}$ is the product of 99 twos and 101 fives, we can pair each of the 99 twos with a five to get 99 tens. Then, we have two 5's left over.

$$2^{99}\times 5^{101}=2^{99}\times 5^{99}\times 5^2$$
$$=(2\times 5)^{99}\times 5^2$$
$$=10^{99}\times 5^2.$$

In order to make a power of 10, we need two more 2's to pair with the remaining two 5's. So, we multiply the numerator and denominator of $\frac{13}{2^{99}\times 5^{101}}$ by 2^2.

$$\frac{13}{2^{99}\times 5^{101}}\overset{\times 2^2}{\underset{\times 2^2}{=}}\frac{13\times 2^2}{2^{101}\times 5^{101}}$$

In the numerator, we have $13\times 2^2=13\times 4=52$.
In the denominator, we have $2^{101}\times 5^{101}=10^{101}$.
So, $\frac{13}{2^{99}\times 5^{101}}=\frac{52}{10^{101}}$.

When $\frac{52}{10^{101}}$ is written as a decimal there are 101 digits to the right of the decimal point: 99 zeros, followed by the digits 5 and 2. So, the sum of the digits to the right of the decimal point is $5+2=\textbf{7}$.

154. To compute $\frac{4}{5}=4\div 5$ using long division, we write 4 as 4.0.

$$5)\overline{4.0}$$

Then, we can ignore the decimal point for a moment and think of $5)\overline{4.0}$ as $5)\overline{40}$.

Since $5\times 8=40$, we know that 5 goes into 40 8 times with 0 left over.

Dividing 5 into 4.0 is similar.

Since $5\times 0.8=4.0$, we know that 5 goes into 4.0 0.8 times with 0 left over.

$$\begin{array}{r} 0.8 \\ 5)\overline{4.0} \\ -4.0 \\ \hline 0 \end{array}$$

So, $\frac{4}{5}=4\div 5=\textbf{0.8}$.

Check: $5\times 0.8=4.0=4.$ ✓

155. To compute $\frac{3}{4}=3\div 4$ using long division, we write 3 as 3.0. Then, we can ignore the decimal point for a moment and think of $4)\overline{3.0}$ as $4)\overline{30}$.

$$4)\overline{3.0}$$

Since $4\times 7=28$, we know 4 goes into 30 a total of 7 times with 2 left over.

Dividing 4 into 3.0 is similar.

Since $4\times 0.7=2.8$, we know 4 goes into 3.0 a total of 0.7 times with 0.2 left over.

$$\begin{array}{r} 0.7 \\ 4)\overline{3.0} \\ -2.8 \\ \hline 0.2 \end{array}$$

Next, we divide 4 into 0.2. This is easier to think about if we write 0.2 as 0.20 and consider dividing 4 into 20.

Since $4\times 5=20$, we know 4 goes into 20 a total of 5 times with 0 left over.

Dividing 4 into 0.20 is similar.

Since $4\times 0.05=0.20$, we know 4 goes into 0.20 a total of 0.05 times with 0 left over.

$$\left.\begin{array}{r} 0.05 \\ 0.7 \\ 4)\overline{3.0} \\ -2.8 \\ \hline 0.20 \\ -0.20 \\ \hline 0 \end{array}\right\}0.75$$

So, $\frac{3}{4}=3\div 4=\textbf{0.75}$.

Check: $4\times 0.75=3.00=3.$ ✓

156. We use the following steps to write $\frac{1}{40}$ as a decimal.

$$40)\overline{1.00}$$

$$\begin{array}{r} 0.02 \\ 40)\overline{1.00} \\ -0.80 \\ \hline 0.20 \end{array}$$

$$\left.\begin{array}{r} 0.005 \\ 0.02 \\ 40)\overline{1.00} \\ -0.80 \\ \hline 0.200 \\ -0.200 \\ \hline 0 \end{array}\right\}0.025$$

So, $\frac{1}{40}=1\div 40=\textbf{0.025}$.

Check: $40\times 0.025=1.00=1.$ ✓

157. We use the following steps to write $\frac{15}{8}$ as a decimal.

$$\begin{array}{r} 1.0 \\ 8)\overline{15.0} \\ -8.0 \\ \hline 7.0 \end{array}$$

$$\begin{array}{r} 0.8 \\ 1.0 \\ 8)\overline{15.0} \\ -8.0 \\ \hline 7.0 \\ -6.4 \\ \hline 0.6 \end{array}$$

$$\begin{array}{r} 0.07 \\ 0.8 \\ 1.0 \\ 8)\overline{15.0} \\ -8.0 \\ \hline 7.0 \\ -6.4 \\ \hline 0.60 \\ -0.56 \\ \hline 0.04 \end{array}$$

$$\left.\begin{array}{r} 0.005 \\ 0.07 \\ 0.8 \\ 1.0 \\ 8)\overline{15.0} \\ -8.0 \\ \hline 7.0 \\ -6.4 \\ \hline 0.60 \\ -0.56 \\ \hline 0.040 \\ -0.040 \\ \hline 0 \end{array}\right\}1.875$$

So, $\frac{15}{8}=15\div 8=\textbf{1.875}$.

— *or* —

We write $\frac{15}{8}$ as $1\frac{7}{8}$, then write $\frac{7}{8}$ as a decimal as shown below.

$$\begin{array}{r} 0.8 \\ 8)\overline{7.0} \\ -6.4 \\ \hline 0.6 \end{array}$$

$$\begin{array}{r} 0.07 \\ 0.8 \\ 8)\overline{7.0} \\ -6.4 \\ \hline 0.60 \\ -0.56 \\ \hline 0.04 \end{array}$$

$$\left.\begin{array}{r} 0.005 \\ 0.07 \\ 0.8 \\ 8)\overline{7.0} \\ -6.4 \\ \hline 0.60 \\ -0.56 \\ \hline 0.040 \\ -0.040 \\ \hline 0 \end{array}\right\}0.875$$

Since $\frac{7}{8}=0.875$, we have $\frac{15}{8}=1\frac{7}{8}=\textbf{1.875}$.

Check: $8\times 1.875=15.000=15.$ ✓

158. We use the following steps to write $\frac{1}{3}$ as a decimal.

$$3)\overline{1.0}$$

$$\begin{array}{r} 0.3 \\ 3)\overline{1.0} \\ -0.9 \\ \hline 0.1 \end{array}$$

$$\begin{array}{r} 0.03 \\ 0.3 \\ 3)\overline{1.0} \\ -0.9 \\ \hline 0.10 \\ -0.09 \\ \hline 0.01 \end{array}$$

$$\begin{array}{r} 0.003 \\ 0.03 \\ 0.3 \\ 3)\overline{1.0} \\ -0.9 \\ \hline 0.10 \\ -0.09 \\ \hline 0.010 \\ -0.009 \\ \hline 0.001 \end{array}$$

We notice a pattern! Each time we divide, we get a remainder that is a decimal ending in 1. When we divide these remainders by 3, the quotient is always a decimal ending in 3.

So, $\frac{1}{3}=0.333...=\textbf{0.}\overline{\textbf{3}}$.

159. We use the following steps to write $\frac{5}{6}$ as a decimal.

$$6)\overline{5.0}$$

$$\begin{array}{r} 0.8 \\ 6)\overline{5.0} \\ -4.8 \\ \hline 0.2 \end{array}$$

$$\begin{array}{r} 0.03 \\ 0.8 \\ 6)\overline{5.0} \\ -4.8 \\ \hline 0.20 \\ -0.18 \\ \hline 0.02 \end{array}$$

$$\begin{array}{r} 0.003 \\ 0.03 \\ 0.8 \\ 6)\overline{5.0} \\ -4.8 \\ \hline 0.20 \\ -0.18 \\ \hline 0.020 \\ -0.018 \\ \hline 0.002 \end{array}$$

After dividing 6 into 5.0, each remainder is a decimal ending in 2. When we divide these remainders by 6, the quotient is a decimal ending in 3. So, every digit after the 8 in our answer is 3.

So, $\frac{5}{6} = 0.8333... = \mathbf{0.8\overline{3}}$.

160. We use the following steps to write $\frac{4}{9}$ as a decimal.

$$9)\overline{4.0}$$

$$\begin{array}{r} 0.4 \\ 9)\overline{4.0} \\ -3.6 \\ \hline 0.4 \end{array}$$

$$\begin{array}{r} 0.04 \\ 0.4 \\ 9)\overline{4.0} \\ -3.6 \\ \hline 0.40 \\ -0.36 \\ \hline 0.04 \end{array}$$

$$\begin{array}{r} 0.004 \\ 0.04 \\ 0.4 \\ 9)\overline{4.0} \\ -3.6 \\ \hline 0.40 \\ -0.36 \\ \hline 0.040 \\ -0.036 \\ \hline 0.004 \end{array}$$

So, $\frac{4}{9} = 0.444... = \mathbf{0.\overline{4}}$.

161. We use the following steps to write $\frac{2}{15}$ as a decimal.

$$15)\overline{2.0}$$

$$\begin{array}{r} 0.1 \\ 15)\overline{2.0} \\ -1.5 \\ \hline 0.5 \end{array}$$

$$\begin{array}{r} 0.03 \\ 0.1 \\ 15)\overline{2.0} \\ -1.5 \\ \hline 0.50 \\ -0.45 \\ \hline 0.05 \end{array}$$

$$\begin{array}{r} 0.003 \\ 0.03 \\ 0.1 \\ 15)\overline{2.0} \\ -1.5 \\ \hline 0.50 \\ -0.45 \\ \hline 0.050 \\ -0.045 \\ \hline 0.005 \end{array}$$

So, $\frac{2}{15} = 0.1333... = \mathbf{0.1\overline{3}}$.

162. We use the following steps to write $\frac{13}{12}$ as a decimal.

$$\begin{array}{r} 1.0 \\ 12)\overline{13.0} \\ -12.0 \\ \hline 1.0 \end{array}$$

$$\begin{array}{r} 0.08 \\ 1.0 \\ 12)\overline{13.0} \\ -12.0 \\ \hline 1.00 \\ -0.96 \\ \hline 0.04 \end{array}$$

$$\begin{array}{r} 0.003 \\ 0.08 \\ 1.0 \\ 12)\overline{13.0} \\ -12.0 \\ \hline 1.00 \\ -0.96 \\ \hline 0.040 \\ -0.036 \\ \hline 0.004 \end{array}$$

$$\begin{array}{r} 0.0003 \\ 0.003 \\ 0.08 \\ 1.0 \\ 12)\overline{13.0} \\ -12.0 \\ \hline 1.00 \\ -0.96 \\ \hline 0.040 \\ -0.036 \\ \hline 0.0040 \\ -0.0036 \\ \hline 0.0004 \end{array}$$

So, $\frac{13}{12} = 1.08333... = \mathbf{1.08\overline{3}}$.

— *or* —

We write $\frac{13}{12}$ as $1\frac{1}{12}$, then write $\frac{1}{12}$ as a decimal.

$$\begin{array}{r} 0.08 \\ 12)\overline{1.00} \\ -0.96 \\ \hline 0.04 \end{array}$$

$$\begin{array}{r} 0.003 \\ 0.08 \\ 12)\overline{1.00} \\ -0.96 \\ \hline 0.040 \\ -0.036 \\ \hline 0.004 \end{array}$$

$$\begin{array}{r} 0.0003 \\ 0.003 \\ 0.08 \\ 12)\overline{1.00} \\ -0.96 \\ \hline 0.040 \\ -0.036 \\ \hline 0.0040 \\ -0.0036 \\ \hline 0.0004 \end{array}$$

Since $\frac{1}{12} = 0.08333... = 0.08\overline{3}$, we have $\frac{13}{12} = 1\frac{1}{12} = \mathbf{1.08\overline{3}}$.

163. First, we write $\frac{3}{11}$ as a decimal using the long division steps shown below.

$$\begin{array}{r} 0.2 \\ 11)\overline{3.0} \\ -2.2 \\ \hline 0.8 \end{array}$$

$$\begin{array}{r} 0.07 \\ 0.2 \\ 11)\overline{3.0} \\ -2.2 \\ \hline 0.80 \\ -0.77 \\ \hline 0.03 \end{array}$$

$$\begin{array}{r} 0.002 \\ 0.07 \\ 0.2 \\ 11)\overline{3.0} \\ -2.2 \\ \hline 0.80 \\ -0.77 \\ \hline 0.030 \\ -0.022 \\ \hline 0.008 \end{array}$$

$$\begin{array}{r} 0.0007 \\ 0.002 \\ 0.07 \\ 0.2 \\ 11)\overline{3.0} \\ -2.2 \\ \hline 0.80 \\ -0.77 \\ \hline 0.030 \\ -0.022 \\ \hline 0.0080 \\ -0.0077 \\ \hline 0.0003 \end{array}$$

The remainders alternate between decimals ending in 8 and 3. When we divide these remainders by 11, the quotients alternate between decimals ending in 2 and 7. So, $\frac{3}{11} = 0.272727... = 0.\overline{27}$.

Every digit that is an even number of places to the right of the decimal point is 7.

So, the 100th digit to the right of the decimal point is **7**.

164. We write $\frac{1}{33}$ as a decimal using the steps shown below.

$$\begin{array}{r} 0.03 \\ 33)\overline{1.00} \\ -0.99 \\ \hline 0.01 \end{array}$$

$$\begin{array}{r} 0.0003 \\ 0.03 \\ 33)\overline{1.00} \\ -0.99 \\ \hline 0.0100 \\ -0.0099 \\ \hline 0.0001 \end{array}$$

$$\begin{array}{r} 0.000003 \\ 0.0003 \\ 0.03 \\ 33)\overline{1.00} \\ -0.99 \\ \hline 0.0100 \\ -0.0099 \\ \hline 0.000100 \\ -0.000099 \\ \hline 0.000001 \end{array}$$

Adding up our quotients, we have

$$\frac{1}{33} = 0.03 + 0.0003 + 0.000003 + \cdots = 0.030303... = 0.\overline{03}.$$

Every digit that is an odd number of places to the right of the decimal point is 0.

So, the 99th digit to the right of the decimal point is **0**.

165. We look for fractions whose denominator can be written as a power of 10.

$\frac{1}{25}$: Multiplying the numerator and denominator by 4, we have $\frac{1}{25} = \frac{4}{100} = 0.04$.

$\frac{1}{30}$: Since $30 = 3 \times 10$, and no power of 10 has a factor of 3, we cannot multiply 30 by an integer to get a power of 10. So, $\frac{1}{30}$ cannot be written as a terminating decimal.

$\frac{1}{91}$: Since $91 = 7 \times 13$, and no power of 10 has a factor of 7 or 13, we cannot write $\frac{1}{91}$ as a fraction with a denominator that is a power of 10. So, $\frac{1}{91}$ cannot be written as a terminating decimal.

$\frac{1}{125}$: Multiplying the numerator and denominator by 8, we have $\frac{1}{125} = \frac{8}{1,000} = 0.008$.

$\frac{1}{55}$: Since $55 = 5 \times 11$, and no power of 10 has a factor of 11, we cannot write $\frac{1}{55}$ as a fraction with a denominator that is a power of 10. So, $\frac{1}{55}$ cannot be written as a terminating decimal.

$\frac{1}{32}$: Multiplying the numerator and denominator by 5^5 gives $\frac{1}{32} = \frac{1}{2^5} = \frac{5^5}{2^5 \times 5^5} = \frac{5^5}{10^5}$. Since $\frac{1}{32}$ can be written as a fraction with a denominator that is a power of 10, it can be written as a terminating decimal. (In fact, $\frac{1}{32} = 0.03125$.)

We circle the fractions that can be written as terminating decimals, as shown below.

166. We're told that $\frac{a}{b}$ can be written as a terminating decimal. So, it is possible to write $\frac{a}{b}$ as an equivalent fraction with a denominator that is a power of 10.

Every power of 10 is a product of 2's and 5's, and no other prime. So, it is possible to multiply b by some integer to get a product of only 2's and 5's.

Therefore, 2 and 5 are the only primes that can be included in the prime factorization of b.

Note that this is always true for fractions in simplest form. If $\frac{a}{b}$ is *not* in simplest form and the prime factorization of b includes primes other than 2 or 5, then it may be possible to write $\frac{a}{b}$ as a terminating decimal.

For example, consider $\frac{a}{b} = \frac{3}{6}$. The prime factorization of b is 2×3. However, the 3 in the prime factorization of a cancels with the 3 in the prime factorization of b when we simplify $\frac{3}{6}$ to $\frac{1}{2}$. So, we can write $\frac{3}{6}$ as the terminating decimal 0.5.

DECIMALS
Conversion Strategies   96-99

167. We notice that $\frac{1}{80} = \frac{1}{10} \times \frac{1}{8}$.

Since $\frac{1}{10} = 0.1$ and $\frac{1}{8} = 0.125$, we have $\frac{1}{80} = 0.1 \times 0.125$.

Multiplying a number by 0.1 moves the decimal point one place to the left. So, $\frac{1}{80} = \frac{1}{10} \times \frac{1}{8} = 0.1 \times 0.125 = \mathbf{0.0125}$.

168. Since $\frac{1}{3} = 0.\overline{3} = 0.333...$, we have $\frac{2}{3} = 2 \times \frac{1}{3} = 2 \times 0.333...$. Multiplying $(0.333...)$ by 2 doubles all of the 3's.

So, $\frac{2}{3} = 2 \times \frac{1}{3} = 2 \times 0.333... = 0.666... = \mathbf{0.\overline{6}}$.

— *or* —

Multiplying a number by 2 is the same as adding 2 copies of that number. So, $2 \times \frac{1}{3} = (0.333...) + (0.333...)$. We stack these decimals vertically, and add them as shown:

$$\begin{array}{r} 0.333... \\ +\,0.333... \\ \hline 0.666... \end{array}$$

So, $\frac{2}{3} = 2 \times \frac{1}{3} = (0.333...) + (0.333...) = 0.666... = \mathbf{0.\overline{6}}$.

169. Since $\frac{1}{3} = 0.\overline{3} = 0.333...$, we have

$$\begin{aligned} \frac{1}{9} &= \frac{1}{3} \times \frac{1}{3} \\ &= \frac{1}{3} \div 3 \\ &= (0.333...) \div 3. \end{aligned}$$

To compute $(0.333...) \div 3$, we consider easier quotients. We know $0.3 \div 3 = 0.1$, $0.33 \div 3 = 0.11$, $0.333 \div 3 = 0.111$, and so on. Continuing the pattern, we have

$$(0.333...) \div 3 = 0.111... = 0.\overline{1}.$$

So, $\frac{1}{9} = \mathbf{0.\overline{1}}$.

170. We use the fact that $\frac{1}{9} = 0.\overline{1}$ to write each fraction as a decimal.

$\frac{2}{9} = 2 \times \frac{1}{9} = 2 \times 0.\overline{1} = 2 \times 0.111... = 0.222... = \mathbf{0.\overline{2}}$.

$\frac{3}{9} = 3 \times \frac{1}{9} = 3 \times 0.\overline{1} = 3 \times 0.111... = 0.333... = \mathbf{0.\overline{3}}$.

$\frac{4}{9} = 4 \times \frac{1}{9} = 4 \times 0.\overline{1} = 4 \times 0.111... = 0.444... = \mathbf{0.\overline{4}}$.

$\frac{5}{9} = 5 \times \frac{1}{9} = 5 \times 0.\overline{1} = 5 \times 0.111... = 0.555... = \mathbf{0.\overline{5}}$.

$\frac{6}{9} = 6 \times \frac{1}{9} = 6 \times 0.\overline{1} = 6 \times 0.111... = 0.666... = \mathbf{0.\overline{6}}$.

$\frac{7}{9} = 7 \times \frac{1}{9} = 7 \times 0.\overline{1} = 7 \times 0.111... = 0.777... = \mathbf{0.\overline{7}}$.

$\frac{8}{9} = 8 \times \frac{1}{9} = 8 \times 0.\overline{1} = 8 \times 0.111... = 0.888... = \mathbf{0.\overline{8}}$.

We notice that for any digit A, the fraction $\frac{A}{9}$ can be written as $0.\overline{A}$.

171. In Problem 169, we found that $\frac{1}{9} = 0.\overline{1}$. So,

$$\frac{1}{90} = \frac{1}{10} \times \frac{1}{9} = 0.1 \times 0.\overline{1} = 0.1 \times 0.111...$$

Multiplying a number by 0.1 moves the decimal point one place to the left. So,

$$\frac{1}{90} = 0.1 \times 0.111... = 0.0111... = 0.0\overline{1}.$$

Then, we have $\frac{1}{45} = \frac{2}{90} = 2 \times \frac{1}{90} = 2 \times 0.0\overline{1} = 0.0\overline{2}$.

Also, we have $\frac{1}{15} = \frac{6}{90} = 6 \times \frac{1}{90} = 6 \times 0.0\overline{1} = 0.0\overline{6}$.

So, the completed statements are as follows:

$\frac{1}{9} = \mathbf{0.\overline{1}}$, so $\frac{1}{90} = \mathbf{0.0\overline{1}}$. Therefore, $\frac{1}{45} = \frac{2}{90} = \mathbf{0.0\overline{2}}$, and $\frac{1}{15} = \frac{6}{90} = \mathbf{0.0\overline{6}}$.

172. Since $\frac{1}{33} = 0.\overline{03} = 0.030303...$, we have

$$\begin{aligned} \frac{1}{99} &= \frac{1}{33} \times \frac{1}{3} \\ &= \frac{1}{33} \div 3 \\ &= (0.030303...) \div 3. \end{aligned}$$

We know $0.03 \div 3 = 0.01$, $0.0303 \div 3 = 0.0101$, and $0.030303 \div 3 = 0.010101$.

Continuing this pattern, we have

$$(0.030303...) \div 3 = 0.010101... = 0.\overline{01}.$$

So, $\frac{1}{99} = \mathbf{0.\overline{01}}$.

173. We use the fact that $\frac{1}{99} = 0.\overline{01}$ to write each fraction as a decimal.

$$\frac{2}{99} = 2 \times \frac{1}{99} = 2 \times 0.\overline{01} = \mathbf{0.\overline{02}}.$$

$$\frac{7}{99} = 7 \times \frac{1}{99} = 7 \times 0.\overline{01} = \mathbf{0.\overline{07}}.$$

$$\frac{17}{99} = 17 \times \frac{1}{99} = 17 \times 0.\overline{01} = \mathbf{0.\overline{17}}.$$

$$\frac{49}{99} = 49 \times \frac{1}{99} = 49 \times 0.\overline{01} = \mathbf{0.\overline{49}}.$$

$$\frac{9}{99} = \frac{1}{11} = 9 \times \frac{1}{99} = 9 \times 0.\overline{01} = \mathbf{0.\overline{09}}.$$

174. From the previous problem, we notice that fractions of the form $\frac{AB}{99}$ can be written as decimals of the form $0.\overline{AB}$, where A and B are digits.

So, we guess that the reverse is true, and that $0.\overline{36} = \frac{36}{99}$. We have

$$\begin{aligned} 0.\overline{36} &= 0.363636... \\ &= 36 \times 0.010101... \\ &= 36 \times 0.\overline{01} \\ &= 36 \times \frac{1}{99} \\ &= \frac{36}{99}. \checkmark \end{aligned}$$

We then simplify our fraction to get $0.\overline{36} = \frac{36}{99} = \frac{4}{11}$.

In general, $\frac{AB}{99} = 0.\overline{AB}$, where A and B are digits.

175. We have $\frac{1}{333} = \frac{1}{3} \times \frac{1}{111} = \frac{1}{3} \div 111 = 0.\overline{333} \div 111$.

To compute $0.\overline{333} \div 111$, we consider easier quotients. $0.333 \div 111 = 0.003$, and $0.333333 \div 111 = 0.003003$. From this pattern, we see that

$$0.\overline{333} \div 111 = 0.\overline{003}.$$

So, $\frac{1}{333} = 0.\overline{003}$.

Then, we have $\frac{1}{999} = \frac{1}{333} \times \frac{1}{3} = \frac{1}{333} \div 3 = 0.\overline{003} \div 3$.

Since $0.003 \div 3 = 0.001$ and $0.003003 \div 3 = 0.001001$, and so on, we see that

$$0.\overline{003} \div 3 = 0.\overline{001}.$$

So, $\frac{1}{999} = 0.\overline{001}$.

Then, $\frac{1}{37} = \frac{27}{999} = 27 \times \frac{1}{999} = 27 \times 0.\overline{001} = 0.\overline{027}$.

So, the completed statements are as follows:

$$\frac{1}{3} = 0.\overline{333}, \text{ so } \frac{1}{333} = \frac{1}{3} \times \frac{1}{111} = \mathbf{0.\overline{003}}, \text{ and } \frac{1}{999} = \mathbf{0.\overline{001}}.$$
$$\text{Therefore, } \frac{1}{37} = \frac{27}{999} = \mathbf{0.\overline{027}}.$$

176. Writing $\frac{100}{99}$ as a mixed number, we have

$$\begin{aligned} \frac{100}{99} &= 1\frac{1}{99} \\ &= 1 + \frac{1}{99} \\ &= 1 + 0.\overline{01} \\ &= \mathbf{1.\overline{01}}. \end{aligned}$$

177. Multiplying the numerator and denominator of $\frac{7}{11}$ by 9, we have $\frac{7}{11} = \frac{63}{99}$.

For any two-digit number AB, we know $\frac{AB}{99} = 0.\overline{AB}$.

So, $\frac{7}{11} = \frac{63}{99} = \mathbf{0.\overline{63}}$.

178. Multiplying the numerator and denominator of $\frac{4}{33}$ by 3, we have $\frac{4}{33} = \frac{12}{99}$.

For any two-digit number AB, we know $\frac{AB}{99} = 0.\overline{AB}$.

So, $\frac{4}{33} = \frac{12}{99} = \mathbf{0.\overline{12}}$.

179. We have $\frac{7}{90} = \frac{7}{9} \times \frac{1}{10} = 0.\overline{7} \times 0.1$.

Multiplying a number by 0.1 moves the decimal point one place to the left. So, $\frac{7}{90} = 0.\overline{7} \times 0.1 = \mathbf{0.0\overline{7}}$.

180. From the previous problem, we know that $0.0\overline{7} = \frac{7}{90}$.

So, $4 \times 0.0\overline{7} = 4 \times \frac{7}{90} = \frac{28}{90}$.

We know that for any digit A, the fraction $\frac{A}{9}$ can be written as $0.\overline{A}$. Using this fact, we have

$$\begin{aligned} \frac{28}{90} &= \frac{28}{9} \times \frac{1}{10} \\ &= 3\frac{1}{9} \times \frac{1}{10} \\ &= 3.\overline{1} \times 0.1 \\ &= 0.3\overline{1}. \end{aligned}$$

So, $4 \times 0.0\overline{7} = \frac{28}{90} = 0.3\overline{1}$.

$0.2\overline{8}$ $0.\overline{28}$ $0.2\overline{9}$ $0.\overline{29}$ $(0.3\overline{1})$ $0.\overline{31}$

181. Multiplying the numerator and denominator of $\frac{13}{18}$ by 5, we have $\frac{13}{18} = \frac{65}{90}$.

We know that for any digit A, the fraction $\frac{A}{9}$ can be written as $0.\overline{A}$. Using this fact, we have

$$\begin{aligned} \frac{65}{90} &= \frac{65}{9} \times \frac{1}{10} \\ &= 7\frac{2}{9} \times \frac{1}{10} \\ &= 7.\overline{2} \times 0.1 \\ &= 0.7\overline{2}. \end{aligned}$$

So, $\frac{13}{18} = \frac{65}{90} = \mathbf{0.7\overline{2}}$.

182. Since $\frac{1}{2} = 0.5$ and $\frac{1}{3} = 0.\overline{3}$, we add $0.5 + 0.\overline{3}$.

We line up the decimal points and write 0's after the 5 in 0.5. Then, we add as shown below.

$$\begin{array}{r} 0.5000... \\ + 0.3333... \\ \hline 0.8333... \end{array}$$

So, $\frac{5}{6} = \frac{1}{2} + \frac{1}{3} = \mathbf{0.8\overline{3}}$.

183. Since $\frac{1}{3} = 0.\overline{3}$, and $\frac{1}{4} = 0.25$, we add $0.\overline{3} + 0.25$.

We line up the decimal points and write 0's after the 5 in 0.25. Then, we add as shown below.

$$\begin{array}{r} 0.33333... \\ + 0.25000... \\ \hline 0.58333... \end{array}$$

So, $\frac{7}{12} = \frac{1}{3} + \frac{1}{4} = \mathbf{0.58\overline{3}}$.

184. Since $\frac{2}{3} = 0.\overline{6}$, and $\frac{1}{4} = 0.25$, we add $0.\overline{6} + 0.25$.

We line up the decimal points and write 0's after the 5 in 0.25. Then, we add as shown below.

$$
\begin{array}{r} \overset{1}{}0.66666... \\ +\ 0.25000... \\ \hline 666... \end{array}
\qquad
\begin{array}{r} \overset{1}{}0.66666... \\ +\ 0.25000... \\ \hline 1666... \end{array}
\qquad
\begin{array}{r} 0.66666... \\ +\ 0.25000... \\ \hline 0.91666... \end{array}
$$

So, $\frac{11}{12} = \frac{2}{3} + \frac{1}{4} = \mathbf{0.91\overline{6}}$.

185. Since $\frac{2}{3} = 0.\overline{6}$, and $\frac{1}{4} = 0.25$, we subtract $0.\overline{6} - 0.25$.

We line up the decimal points and write 0's after the 5 in 0.25. Then, we subtract as shown below.

$$
\begin{array}{r} 0.66666... \\ -\ 0.25000... \\ \hline 0.41666... \end{array}
$$

So, $\frac{5}{12} = \frac{2}{3} - \frac{1}{4} = \mathbf{0.41\overline{6}}$.

186. Since $\frac{1}{3} = 0.\overline{3}$, and $\frac{1}{4} = 0.25$, we subtract $0.\overline{3} - 0.25$.

We line up the decimal points and write 0's after the 5 in 0.25. To subtract, we need to take a tenth from the tenths place in $0.\overline{3}$ to make 13 hundredths. Then, we subtract as shown below.

$$
\begin{array}{r} 0.33333... \\ -\ 0.25000... \\ \hline \end{array}
\qquad
\begin{array}{r} 0.\overset{2}{\cancel{3}}\overset{13}{\cancel{3}}333... \\ -\ 0.25000... \\ \hline \end{array}
\qquad
\begin{array}{r} 0.\overset{2}{\cancel{3}}\overset{13}{\cancel{3}}333... \\ -\ 0.25000... \\ \hline 0.08333... \end{array}
$$

So, $\frac{1}{12} = \frac{1}{3} - \frac{1}{4} = \mathbf{0.08\overline{3}}$.

DECIMALS

Frac-Turns 100-103

187. We begin by converting each fraction to a decimal. Since leading zeros to the left of the decimal point are ignored in these puzzles, we do not write leading zeros when we convert.

$$\frac{1}{4} = .25, \qquad \frac{3}{5} = .6, \qquad \frac{2}{9} = .\overline{2}.$$

There are only two empty squares in the path for $\frac{1}{4} = .25$. So, we fill these squares with 2 and 5, as shown.

Placing the 2 also completes the path for $\frac{2}{9} = .\overline{2}$.

Then, the path for $\frac{3}{5} = .6$ has only one empty square after the decimal point. So, we fill this square with 6.

Each fraction's path correctly traces its decimal form, so we are done.

188. We begin by converting each fraction to a decimal.

$$\frac{8}{3} = 2\frac{2}{3} = 2.\overline{6}$$

$$\frac{1}{6} = \frac{1}{2} - \frac{1}{3} = .5 - .\overline{3} = .1\overline{6}$$

$$\frac{3}{50} = \frac{6}{100} = .06$$

$$\frac{1}{100} = .01$$

There are only two empty squares in the path for $\frac{1}{100} = .01$. So, we fill these squares with 0 and 1, as shown.

Then, to complete the path for $\frac{1}{6} = .1\overline{6}$, we place a 6 in the bottom-middle square.

Placing the 6 also completes the path for $\frac{3}{50} = .06$.

Finally, to complete the path for $\frac{8}{3} = 2.\overline{6}$, we place a 2 in the empty square before the decimal point.

Each fraction's path traces its decimal form, so we are done.

We use the strategies discussed in the previous problems to solve the following puzzles.

189. $\frac{3}{5} = .6$

$\frac{2}{3} = .\overline{6}$

$\frac{1}{6} = .1\overline{6}$

$\frac{1}{10} = .1$

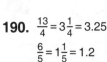

190. $\frac{13}{4} = 3\frac{1}{4} = 3.25$

$\frac{6}{5} = 1\frac{1}{5} = 1.2$

$\frac{21}{4} = 5\frac{1}{4} = 5.25$

191. $\frac{65}{9} = 7\frac{2}{9} = 7.\overline{2}$

$\frac{1}{2} = .5$

$\frac{9}{5} = 1\frac{4}{5} = 1.8$

$\frac{3}{20} = \frac{15}{100} = .15$

$\frac{1}{4} = .25$

192. $\frac{23}{3} = 7\frac{2}{3} = 7.\overline{6}$

$\frac{9}{25} = \frac{36}{100} = .36$

$\frac{35}{12} = 2\frac{11}{12} = 2 + \frac{2}{3} + \frac{1}{4} = 2 + .\overline{6} + .25 = 2.91\overline{6}$

$\frac{11}{30} = \frac{33}{90} = \frac{33}{9} \times \frac{1}{10} = 3\frac{6}{9} \times \frac{1}{10} = 3.\overline{6} \times .1 = .3\overline{6}$

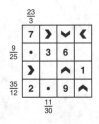

193. We convert each fraction to a decimal.

$$\frac{1}{6} = .1\overline{6}$$

$$\frac{11}{3} = 3\frac{2}{3} = 3.\overline{6}$$

$$\frac{31}{10} = 3\frac{1}{10} = 3.1$$

$$\frac{55}{9} = 6\frac{1}{9} = 6.\overline{1}$$

$$\frac{9}{5} = 1\frac{4}{5} = 1.8$$

Since the squares in the bottom row make up the repeating part of the path for $\frac{1}{6} = .1\overline{6}$, each square must remain empty or contain a 6.

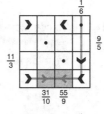

The left empty square is the first square in the path of $\frac{31}{10} = 3.1$, so it cannot be 6. Therefore, it must remain empty. We cross out this square as a reminder that it is empty, then place a 6 in the right square as shown.

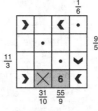

Then, we use the strategies discussed in previous problems to complete the puzzle as shown.

194. We convert each fraction to a decimal.

$$\frac{4}{15} = \frac{24}{90} = \frac{24}{9} \times \frac{1}{10} = 2\frac{6}{9} \times \frac{1}{10} = 2.\overline{6} \times .1 = .2\overline{6}$$

$$\frac{7}{3} = 2\frac{1}{3} = 2.\overline{3}$$

$$\frac{9}{2} = 4\frac{1}{2} = 4.5$$

$$\frac{3}{10} = .3$$

We consider the squares in the path of $\frac{4}{15} = .2\overline{6}$, each of which must be empty or filled with a 2 or a 6.

One of these squares is the first square in the path of $\frac{9}{2} = 4.5$, which cannot have any 2's or 6's.

Another square is the first square in the path of $\frac{3}{10} = .3$, which cannot have any 2's or 6's.

So, we place an ✕ in these two squares, then fill the remaining empty squares in the path of $\frac{4}{15} = .2\overline{6}$ with 2 and 6 as shown.

Then, we use the strategies discussed in previous problems to complete the puzzle as shown.

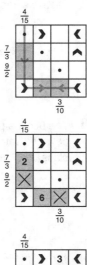

195. We convert each fraction to a decimal.

$$\frac{4}{5} = .8$$

$$\frac{9}{50} = \frac{18}{100} = .18$$

$$\frac{57}{11} = 5\frac{2}{11} = 5\frac{18}{99} = 5.\overline{18}$$

$$\frac{22}{25} = \frac{88}{100} = .88$$

$$\frac{9}{11} = \frac{81}{99} = .\overline{81}$$

$$\frac{2}{11} = \frac{18}{99} = .\overline{18}$$

In the second row of the grid, there are three empty squares which are part of the path of $\frac{4}{5} = .8$. So, exactly one of these squares contains an 8.

The path of $\frac{22}{25} = .88$ crosses an empty square in the third row. Then, it crosses the three empty squares in the second row, where there can only be one 8. So, we must place an 8 in the third row.

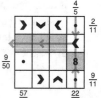

Then, we consider the loop that is shared by the paths for $\frac{57}{11} = 5.\overline{18}$, $\frac{9}{11} = .\overline{81}$, and $\frac{2}{11} = .\overline{18}$. One of the squares in this loop contains a 1, and another contains an 8.

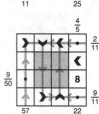

1 cannot go in either of the loop's top squares, since they are in the path of $\frac{4}{5} = .8$. Also, 1 cannot go in the loop's bottom-right square, since this is the first empty square in the path of $\frac{9}{11} = .\overline{81}$. So, 1 must go in the bottom-left square. This completes the path for $\frac{9}{50} = .18$, so we place an ✕ to the right of the 1.

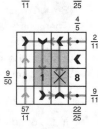

Then, we cannot place an 8 above the 1, since this is the first empty square in the path of $\frac{2}{11} = .\overline{18}$. So, we must place the 8 as shown.

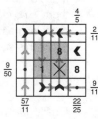

Finally, we place a 5 in the first square in the path of $\frac{57}{11} = 5.\overline{18}$.

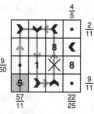

Each fraction's path correctly traces its decimal form, so we are done.

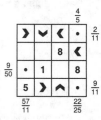

196. We have $\frac{1}{4} = .25$ and $\frac{5}{2} = 2.5$.

So, the first square in the path of $\frac{5}{2}$ must be filled with 2, and the first square in the path of the bottom $\frac{1}{4}$ must be empty.

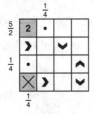

Then, we consider the path of the bottom $\frac{1}{4}$.

Exactly one of the two shaded squares shown to the right must contain a 2.

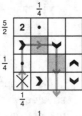

If we place the 2 as shown, then we must fill the remaining empty squares in the bottom $\frac{1}{4}$'s path with an $\times$ and a 5.

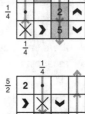

Also, the square to the left of the 2 must be filled with a 2 to complete the path of the top $\frac{1}{4}$.

However, this forces the first two digits in the left $\frac{1}{4}$'s path to be 2's. ✗

So, we must place a 2 as shown. Then, there is no digit that can go beneath this 2 that will satisfy the paths of the left $\frac{1}{4}$ *and* the top $\frac{1}{4}$. So, this square must be empty.

In the shaded square on the right, there is no digit that can satisfy the paths of the left $\frac{1}{4}$ *and* the bottom $\frac{1}{4}$. So, this square must also be empty.

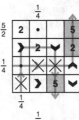

Finally, the last square in the path of the bottom $\frac{1}{4}$ must be 5, and the last two squares in the path of the left $\frac{1}{4}$ must be 2 and 5.

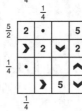

Each fraction's path correctly traces its decimal form, so we are done.

197. We complete the paths for $\frac{1}{5} = .2$, $\frac{11}{20} = .55$, and $\frac{9}{4} = 2.25$.

Completing the path for $\frac{9}{4} = 2.25$ also completes the path for $\frac{21}{4} = 5.25$.

Then, we complete the path for $\frac{11}{2} = 5.5$. This completes the path for $\frac{13}{25} = .52$.

Finally, we place the 5 in the path for $\frac{1}{2} = .5$ in the only square that does not interfere with another path.

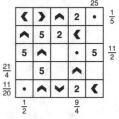

198. We complete the path for $\frac{7}{4} = 1.75$, placing the 1 so that it is not in the path of $\frac{53}{10} = 5.3$.

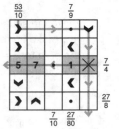

Then, we complete the path for $\frac{27}{8} = 3.375$, placing the second 3 so that it is not in the path of $\frac{7}{10} = .7$, and placing the 5 so that it is not in the path of $\frac{7}{9} = .\overline{7}$.

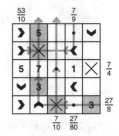

This also completes the paths for $\frac{53}{10} = 5.3$ and $\frac{27}{80} = .3375$.

Then, we complete the path for $\frac{7}{9} = .\overline{7}$.

This also completes the path for $\frac{7}{10} = .7$.

Each fraction's path correctly traces its decimal form, so we are done.

199. The 5 in the path of $\frac{5}{9} = .\overline{5}$ cannot be in the path of $\frac{7}{25} = .28$, so we place the 5 as shown.

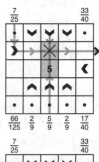

Then, we complete the path for $\frac{17}{40} = .425$. This completes the path for the $\frac{2}{9} = .\overline{2}$ on the right.

We place $\times$'s in the squares at the ends of these two paths as shown.

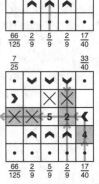

Then, we complete the path for the $\frac{2}{9} = .\overline{2}$ on the left, then the path for $\frac{7}{25} = .28$. This also completes the path for $\frac{33}{40} = .825$.

Finally, we place a 5 to complete the path for $\frac{66}{125} = .528$.

Each fraction's path correctly traces its decimal form, so we are done.

200. The empty squares in the middle row are in the repeating part of the paths for $\frac{31}{9} = 3.\overline{4}$, $\frac{29}{45} = .6\overline{4}$, and $\frac{4}{9} = .\overline{4}$. So, each of these squares must remain empty or contain a 4.

We consider the loop that is shared by the following paths: $\frac{6}{11} = .\overline{54}$, $\frac{17}{11} = 1.\overline{54}$, $\frac{5}{11} = .\overline{45}$. One of the squares in this loop contains a 5, and another contains a 4.

5 cannot go in either of the loop's top squares, since the only digit that can go in either square is 4. Also, 5 cannot go in the loop's lower-left square, since this is the first empty square in the path of $\frac{5}{11} = .\overline{45}$. So, a 5 must go in the loop's lower-right square.

We cannot place a 4 above the 5, since this is the first empty square in the path of $\frac{6}{11} = .\overline{54}$. Also, we cannot place a 4 to the left of the 5 because this square is in the path of $\frac{32}{125} = .256$.

So, we place a 4 in the loop's upper-left square.

The paths for $\frac{4}{9} = .\overline{4}$, $\frac{5}{11} = .\overline{45}$, and $\frac{6}{11} = .\overline{54}$ are now complete.

We finish the puzzle by completing the following paths:

$\frac{29}{45} = .6\overline{4}$, $\frac{32}{125} = .256$, $\frac{2}{9} = .\overline{2}$,

$\frac{17}{11} = 1.\overline{54}$, $\frac{31}{9} = 3.\overline{4}$.

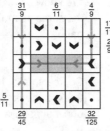

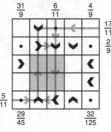

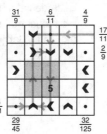

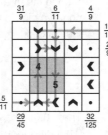

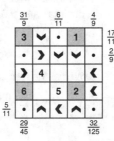

201. We convert each of the given fractions to decimals.

$\frac{29}{40} = .725$ $\qquad$ $\frac{1}{6} = .1\overline{6}$

$\frac{21}{4} = 5.25$ $\qquad$ $\frac{2}{11} = .\overline{18}$

$\frac{9}{4} = 2.25$ $\qquad$ $\frac{9}{11} = .\overline{81}$

$\frac{9}{10} = .9$ $\qquad$ $\frac{1}{15} = .0\overline{6}$

$\frac{24}{5} = 4.8$

There is only one way to place these fractions and exactly one of each digit so that each fraction's path correctly traces its decimal form. *(For a complete step-by-step solution, visit www.BeastAcademy.com.)*

Dividing Decimals $\qquad$ 104

202. We write $0.05 \div 1.25$ as $\frac{0.05}{1.25}$. Then, we multiply the numerator and denominator of $\frac{0.05}{1.25}$ by 100 to get an equivalent fraction whose numerator and denominator are both integers.

$$\frac{0.05}{1.25} \xrightarrow[\times 100]{\times 100} = \frac{5}{125}$$

$\frac{5}{125}$ simplifies to $\frac{1}{25}$, so we have $0.05 \div 1.25 = \mathbf{\frac{1}{25}}$.

203. We write $0.4 \div 2.2$ as $\frac{0.4}{2.2}$. Then, we multiply the numerator and denominator of $\frac{0.4}{2.2}$ by 10 to get an equivalent fraction whose numerator and denominator are both integers.

$$\frac{0.4}{2.2} \xrightarrow[\times 10]{\times 10} = \frac{4}{22}$$

$\frac{4}{22}$ simplifies to $\frac{2}{11}$, so we have $0.4 \div 2.2 = \mathbf{\frac{2}{11}}$.

204. We write $3.03 \div 0.6$ as $\frac{3.03}{0.6}$. Then, multiplying the numerator and denominator of $\frac{3.03}{0.6}$ by 100 gives

$$\frac{3.03}{0.6} = \frac{303}{60} = \frac{101}{20} = \mathbf{5\frac{1}{20}}.$$

205. We write $3.5 \div 70.707$ as $\frac{3.5}{70.707}$. Then, multiplying the numerator and denominator of $\frac{3.5}{70.707}$ by 1,000 gives

$$\frac{3.5}{70.707} = \frac{3,500}{70,707} = \mathbf{\frac{500}{10,101}}.$$

206. Multiplying the numerator and denominator of $\frac{0.35}{0.05}$ by 100 gives

$$\frac{0.35}{0.05} = \frac{35}{5} = \textbf{7}.$$

207. Multiplying the numerator and denominator of $\frac{2.4}{0.03}$ by 100 gives

$$\frac{2.4}{0.03} = \frac{240}{3} = \textbf{80}.$$

208. Multiplying the numerator and denominator of $\frac{0.032}{0.8}$ by 1,000 gives

$$\frac{0.032}{0.8} = \frac{32}{800} = \frac{4}{100} = \textbf{0.04}.$$

209. We write $4.9 \div 0.14$ as a quotient of integers, then divide:

$$4.9 \div 0.14 = \frac{4.9}{0.14} = \frac{490}{14} = \textbf{35}.$$

210. We write $0.002 \div 0.04$ as a quotient of integers, then convert the resulting fraction to a decimal:

$$0.002 \div 0.04 = \frac{0.002}{0.04} = \frac{2}{40} = \frac{1}{20} = \frac{5}{100} = \textbf{0.05}.$$

211. We write $0.6 \div 4.5$ as a quotient of integers, then convert the resulting fraction to a decimal:

$$0.6 \div 4.5 = \frac{0.6}{4.5} = \frac{6}{45} = \frac{12}{90} = \frac{12}{9} \times \frac{1}{10} = 1\frac{1}{3} \times 0.1 = \textbf{0.1}\overline{\textbf{3}}.$$

Challenge Problems   106-107

212. Since $\frac{1}{2} = 0.5$, we have $\left(\frac{1}{2}\right)^{99} = (0.5)^{99}$. To find the rightmost digit of $(0.5)^{99}$, we temporarily ignore the decimal point and consider 5^{99}.

All positive powers of 5 have units digit 5. For example, $5^1 = \underline{5}$, $5^2 = 2\underline{5}$, $5^3 = 12\underline{5}$, and so on. So, 5^{99} is equal to some integer whose rightmost digit is 5.

To compute $(0.5)^{99}$, we move the decimal point in 5^{99} ninety-nine places to the left. Moving the decimal point left does not change the rightmost digit. So, the rightmost digit of $(0.5)^{99}$ is **5**.

213. Since $\frac{1}{5} = 0.2$, we have $\left(\frac{1}{5}\right)^{10} = (0.2)^{10}$.

To compute $(0.2)^{10}$, we move the decimal point in $2^{10} = 1{,}024$ ten places to the left.

Since 1,024 has 4 digits, moving the decimal point ten places to the left gives $10 - 4 = \textbf{6}$ zeros to the right of the decimal point before the first nonzero digit.

We have

$$\left(\frac{1}{5}\right)^{10} = (0.2)^{10} = 0.\underset{\sim}{0000001024}.$$

214. Since potatoes cost $0.67 per pound, the cost of 6.375 pounds of potatoes is 0.67×6.375 dollars.

Since 0.67 is less than 0.7 and 6.375 is less than 7, we know that 0.67×6.375 is less than $0.7 \times 7 = 4.9$.

Since 0.67 is more than $0.\overline{6} = \frac{2}{3}$ and 6.375 is more than 6, we know that 0.67×6.375 is more than $\frac{2}{3} \times 6 = 4$.

So, 0.67×6.375 is between 4 and 4.9. Therefore, the smallest whole number of dollars needed to buy the potatoes is **5**.

In fact, $0.67 \times 6.375 = 4.27125$.

215. The smallest positive number with three digits after the decimal point is 0.001. So, we look for the number that makes the following equation true:

$$0.008 \times \boxed{} = 0.001.$$

If $0.008 \times \boxed{} = 0.001$, then $0.001 \div 0.008 = \boxed{}$.

We can write $0.001 \div 0.008$ as $\frac{0.001}{0.008}$, then multiply the numerator and denominator by 1,000 to get $\frac{0.001}{0.008} = \frac{1}{8}$.

So, the smallest positive number we can multiply by 0.008 to get a decimal product with 3 digits after the decimal point is $\frac{1}{8} = \textbf{0.125}$.

— *or* —

The smallest positive number with three digits after the decimal point is 0.001.

We know $0.008 \div 8 = 0.001$. Since dividing by a number is the same as multiplying by that number's reciprocal, we have

$$0.008 \times \frac{1}{8} = 0.001.$$

So, the smallest positive number we can multiply by 0.008 to get a decimal product with 3 digits after the decimal point is $\frac{1}{8} = \textbf{0.125}$.

216. To make our comparisons easier, we write out the first several digits of each decimal. Then, we line up our place values, filling any empty place values with trailing zeros.

$$0.\overline{05} = 0.050505...$$
$$0.\overline{050} = 0.050050...$$
$$0.05 = 0.050000$$
$$0.0\overline{5} = 0.055555...$$

Each decimal has 0 ones, 0 tenths, and 5 hundredths. In the thousandths place, only $0.0\overline{5}$ has a nonzero digit. So, $0.0\overline{5}$ is the largest decimal.

$$0.\overline{05} = 0.05\boxed{0}505...$$
$$0.\overline{050} = 0.05\boxed{0}050...$$
$$0.05 = 0.05\boxed{0}000$$
$$0.0\overline{5} = 0.05\boxed{5}555...$$

Among the three remaining decimals, only $0.\overline{05}$ has a nonzero digit in the ten-thousandths place. So, $0.\overline{05}$ is the second-largest decimal.

$$0.\overline{05} = 0.050\boxed{5}05...$$
$$0.\overline{050} = 0.050\boxed{0}50...$$
$$0.05 = 0.050\boxed{0}00$$

Then, $0.\overline{050}$ has 5 hundred-thousandths while 0.05 has 0 hundred-thousandths. So, $0.\overline{050}$ is the third-largest decimal, and 0.05 is the smallest.

$$0.\overline{050} = 0.0500\boxed{5}0...$$
$$0.05 = 0.0500\boxed{0}0$$

We list the four decimals in order from least to greatest.

$$\textbf{0.05,} \quad \textbf{0.}\overline{\textbf{050}}\textbf{,} \quad \textbf{0.}\overline{\textbf{05}}\textbf{,} \quad \textbf{0.0}\overline{\textbf{5}}$$

217. We begin by writing each term in the sum as a fraction.

$$0.\overline{1} = \frac{1}{9}.$$

$$0.0\overline{1} = 0.\overline{1} \times 0.1 = \frac{1}{9} \times \frac{1}{10} = \frac{1}{90}.$$

$$0.00\overline{1} = 0.0\overline{1} \times 0.1 = \frac{1}{90} \times \frac{1}{10} = \frac{1}{900}.$$

So, $0.\overline{1} + 0.0\overline{1} + 0.00\overline{1} = \frac{1}{9} + \frac{1}{90} + \frac{1}{900}$.

We write each fraction with a common denominator of 900, then add:

$$\frac{1}{9} + \frac{1}{90} + \frac{1}{900} = \frac{100}{900} + \frac{10}{900} + \frac{1}{900}$$
$$= \frac{111}{900}$$
$$= \frac{37}{300}.$$

So, $0.\overline{1} + 0.0\overline{1} + 0.00\overline{1} = \frac{37}{300}$.

— *or* —

We add the decimals vertically, writing the first several digits to the right of each decimal point.

$$
\begin{array}{r}
0.11111... \\
0.01111... \\
+0.00111... \\
\hline
0.12333...
\end{array}
$$

So, $0.\overline{1} + 0.0\overline{1} + 0.00\overline{1} = 0.12\overline{3}$. We use the steps shown below to write $0.12\overline{3}$ as a fraction:

$$0.12\overline{3} = 12.\overline{3} \times 0.01$$
$$= 12\frac{1}{3} \times \frac{1}{100}$$
$$= \frac{37}{3} \times \frac{1}{100}$$
$$= \frac{37}{300}.$$

218. Only a fraction that can be written with a denominator that is a power of 10 can be written as a terminating decimal. So, if a fraction can be written as a terminating decimal, then 2's and 5's are the only primes in the prime factorization of the denominator.

The denominator of $\frac{1}{8!}$ is 8!, and the prime factorization of 8! is

$$8! = 8 \times 7 \times 6 \times 5 \times 4 \times 3 \times 2 \times 1$$
$$= (2 \times 2 \times 2) \times 7 \times (2 \times 3) \times 5 \times (2 \times 2) \times 3 \times 2$$
$$= 2^7 \times 3^2 \times 5 \times 7.$$

In order to write the denominator of $\frac{1}{8!}$ as a power of 10, we must cancel the factors of 3^2 and 7 in (8!).

The smallest number that has 3^2 and 7 as factors is $3^2 \times 7 = 9 \times 7 = 63$. Multiplying $\frac{1}{8!}$ by 63 gives

$$\frac{1}{8!} \times 63 = \frac{63}{8!}$$
$$= \frac{\overset{1}{\cancel{3^2}} \times \overset{1}{\cancel{7}}}{2^7 \times \underset{1}{\cancel{3^2}} \times 5 \times \underset{1}{\cancel{7}}}$$
$$= \frac{1}{2^7 \times 5}.$$

Since $\frac{1}{2^7 \times 5}$ has only factors of 2 and 5 in the denominator, it can be written as an equivalent fraction with a denominator that is a power of 10.

$$\frac{1}{2^7 \times 5} = \frac{1}{2^7 \times 5} \times \frac{5^6}{5^6}$$
$$= \frac{5^6}{2^7 \times 5^7}$$
$$= \frac{5^6}{10^7}.$$

Multiplying $\frac{1}{8!}$ by any number smaller than 63 leaves a prime factor other than 2 or 5 in the denominator of the resulting fraction.

So, **63** is the smallest positive integer that can be multiplied by $\frac{1}{8!}$ to get a result that can be expressed as a terminating decimal.

219. In order to divide, we write each decimal as a fraction.

$$0.0\overline{3} = 0.\overline{3} \times 0.1 = \frac{1}{3} \times \frac{1}{10} = \frac{1}{30}.$$
$$0.\overline{2} = \frac{2}{9}.$$

So, we have

$$\frac{0.0\overline{3}}{0.\overline{2}} = 0.0\overline{3} \div 0.\overline{2} = \frac{1}{30} \div \frac{2}{9}.$$

To divide by a number, we multiply by its reciprocal. So,

$$\frac{1}{30} \div \frac{2}{9} = \frac{1}{30} \times \frac{9}{2} = \frac{9}{60} = \frac{3}{20}.$$

Finally, we convert $\frac{3}{20}$ to a decimal:

$$\frac{3}{20} = \frac{15}{100} = 0.15.$$

So, $\frac{0.0\overline{3}}{0.\overline{2}} = \textbf{0.15}$.

220. **Since $0.\overline{3} = \frac{1}{3}$, adding three copies of $0.\overline{3}$ is the same as adding three copies of $\frac{1}{3}$. So, Grogg's sum is equal to**

$$\frac{1}{3} + \frac{1}{3} + \frac{1}{3} = \frac{3}{3} = \textbf{1}.$$

We have just shown that $0.\overline{9}$ and 1 are the same number! This may sound strange at first, but one way to convince ourselves that this is true is by trying to find a number between $0.\overline{9}$ and 1.

If two numbers are different, then they fall on different places on the number line, and we can always zoom in and find some number in between them on the number line. However, there are infinitely many 9's after the decimal point in $0.\overline{9}$, so there is no number we can write that is greater than $0.\overline{9}$ while also being less than 1. (Try it!). Since there is no number between $0.\overline{9}$ and 1, they must be the same number!

If this is still confusing to you, don't worry too much. Things involving infinity can be very tough to think about! As you continue learning math, you'll discover other ways to show that $0.\overline{9}$ and 1 are the same number.